PERGAMON INTERNATIONAL LIBRARY
of Science, Technology, Engineering and Social Studies

The 1000-volume original paperback library in aid of education, industrial training and the enjoyment of leisure

Publisher: Robert Maxwell, M.C

MATRIX METHODS OF STRUCTURAL ANALYSIS

Other Titles of Interest in the

PERGAMON INTERNATIONAL LIBRARY

Browne:
Basic Theory of Structures

Charlton:
Model Analysis of Flame Structures

Davies:
Structural Concrete

Dugdale:
Elements of Elasticity

Dym:
Introduction to the Theory of Shells

Hendry:
Elements of Experimental Stress Analysis

Horne and Merchant:
The Stability of Frames

Lenczner:
The Elements of Load-bearing Brickwork

Lenczner:
Movements in Buildings

Neal:
Structural Theorems and their Applications

Parkes:
Braced Frameworks, 2nd Edition

Whitaker:
The Design of Piled Foundations.

Matrix Methods of Structural Analysis

2ND EDITION

R. K. LIVESLEY
University Lecturer in Engineering,
Fellow of Churchill College, Cambridge

PERGAMON PRESS
OXFORD · NEW YORK · TORONTO
SYDNEY · BRAUNSCHWEIG

U.K.	Pergamon Press Ltd., Headington Hill Hall, Oxford OX3 0BW, England
U.S.A.	Pergamon Press Inc., Maxwell House, Fairview Park, Elmsford, New York 10523, U.S.A.
CANADA	Pergamon of Canada, Ltd., 207 Queen's Quay West, Toronto 1, Canada
AUSTRALIA	Pergamon Press (Aust.) Pty. Ltd., 19a Boundary Street, Rushcutters Bay, N.S.W. 2011, Australia
FRANCE	Pergamon Press SARL, 24 rue des Ecoles, 75240 Paris, Cedex 05, France
WEST GERMANY	Pergamon Press GmbH, 3300 Braunschweig, Postfach 2923, Burgplatz 1, West Germany

First edition 1964

Second edition 1975

Library of Congress Cataloging in Publication Data

Livesley, R K
Matrix methods of structural analysis

(Structures and solid body mechanics division)
Bibliography: p.
1. Structures, Theory of--Matrix methods. I. Title
TA642.L5 1975 624'. 171'01512943 75-12835
ISBN 0-08-018888-5
ISBN 0-08-018887-7 flexicover

Printed in Great Britain by A. Wheaton & Co., Exeter

Contents

Preface to the Second Edition

A great deal has happened to the theory and practice of structural analysis since the first edition of this book was written. New theoretical ideas have increased the range of problems which can be solved by matrix methods. The development of large suites of computer programs such as the ICES and GENESYS systems has made these theoretical advances available to the non-specialist practising engineer. Most important of all, a link to a computer is now accepted as a normal item of equipment in a structural design office. As a consequence, matrix methods are now regarded as the standard tools for solving most non-trivial problems of structural analysis.

The arrangement of the new edition reflects some of these changes. While the book retains its emphasis on skeletal structures (still of prime importance to the majority of structural engineers) the presentation now follows a more general finite-element approach. The chapter on line elements (i.e. beams, columns and arches) is followed by a parallel chapter on some of the simpler types of element used in continuum analysis. A new chapter on rigid-plastic methods for the analysis and design of frameworks is included. This is based on material presented to a conference on optimum design at Swansea in 1972. The methods presented in this chapter have natural links with techniques for automatic redundant selection in elastic analysis, and one such technique is now included in the account of the compatibility method. The sections on vibration analysis have also been extended to include an account of consistent mass matrices.

To make room for the new topics I have omitted some theoretical material which the passage of time has shown to be of little value or interest. I have also omitted most of the original discussion of general computing principles, since there can now be few engineers or students of engineering without some experience of using a computer. The chapter on matrix algebra has been relegated to an appendix, since the material it contains is now covered in most engineering mathematics courses.

Although the new material increases the range of subject matter treated, I have made no attempt to cover all the developments in structural matrix theory and computing techniques which have taken place in the last 10 years. There are now a number of comprehensive texts in which the reader will find extensions and refinements of the methods presented here. In the treatment of continuum problems, particularly, this book attempts no more than a simple introduction to a rapidly growing area of specialization. As in the earlier edition, the general level of structural and mathematical knowledge assumed is roughly that appropriate to a final-year degree course in civil engineering.

As the book is intended for a newcomer to the subject I have tried to emphasize those ideas which are of central importance to structural matrix theory. I have chosen to place particular emphasis on two ideas which seem to me to be important components of any general systematic theory of structures.

Not surprisingly, the first of these is the principle of virtual work. This principle can be applied to elastic and inelastic materials, to line, area and volume elements, to linear and non-linear structures. It provides an essential link between equations of equilibrium and equations of compatibility, and the reader will find examples of its use throughout the book.

The second idea is a very simple principle, which is not followed in many textbooks and is rarely stated explicitly even when it is used. The principle is simply that the definitive equations which specify the load/displacement characteristics of any structural element should be expressed in a form which is (a) independent of rigid-body displacements of the element and (b) automatically satisfies the equations of overall element equilibrium. To achieve this I have introduced the concepts of the element deformation vector $\boldsymbol{e}$ and the element stress-resultant vector $\boldsymbol{r}$, using these concepts consistently to provide a unified approach to the analysis of different types of element.

Although the use of the vectors $\boldsymbol{r}$ and $\boldsymbol{e}$ in the new edition seems to me to clarify the presentation, those readers familiar with the earlier edition of the book should note that it has involved some changes in notation. The element equilibrium matrices $\boldsymbol{H}_1$, $\boldsymbol{H}_2$ are now defined in a slightly different way, while the symbol $\mathbf{H}$ is used to represent the overall structural equilibrium matrix previously denoted by **C**. I hope that these changes will not cause confusion.

When teaching a traditional manual method of structural analysis it is taken for granted that the person taught needs both a general understanding of the method and practice in applying it to specific problems. Most structural textbooks therefore contain examples worked in the text and other problems for solution by the reader. With matrix methods the situation is different. The aim of matrix theory is to set up a formal procedure which can form the basis of a computer program. Obviously the writer of such a program should "understand" the procedure, but he need not be an expert at carrying out the standard operations of matrix algebra by hand. Indeed, it is likely that in his programming he will use library sub-routines for these operations. It does not seem to me that the manual working of numerical examples is an appropriate way of generating the sort of understanding required in this situation. Consequently I have not included the usual problems for solution by the reader.

The generality of matrix methods links up with another aspect of digital computing which also raises a problem for the teacher of structural theory. The time and expense associated with writing computer programs for structural analysis has brought about a situation where there are a few expert "program writers" and many non-specialist "program users". The program writer clearly needs a detailed understanding of the relevant structural theory, although in the development of large general-purpose systems the bulk of the work is likely to be concerned more with data handling than analysis. But what should the program user know? He, after all, will get correct results from the program (if his data is correct) whether he understands the theory or not.

I do not myself see a clear answer to this question. As a program writer I naturally hope that this book will be useful to readers doing similar work who require details of formulae and procedures. However, I also hope that program users will find in it a general account of structural theory which will help them to use existing computer programs in a sensible manner.

When the first edition was written, most of the work which it described could be labelled "recent research" and it was appropriate to give references to the original papers in which the work was first described. With the passage of time it seems less appropriate to direct readers to the original sources, since most of the material is now available in books such as Przemieniecki (1968).† (This book also contains an excellent bibliography.) However, I have retained references to a few of the earlier papers which have historic or personal significance.

In the preface to the first edition I acknowledged my debt to those who encouraged my research work in the early days of structural computing. While that debt still stands, there are a number of other people who have contributed ideas or material to the new edition. In particular, I thank A. P. Kfouri for helpful discussions in connection with the material in Chapter 4 and Professor O. C. Zienkiewicz for originally inviting me to prepare the material which forms the basis for Chapter 7. The computer program which produced Fig. 10.10 was developed as a final-year project by S. P. Marchand, an undergraduate in the Cambridge University Engineering Department.

It is not only the ideas presented in this book which are linked with computers. The manuscript was put onto paper tape at an early stage and much of the subsequent editing was done on a computer. I am grateful to the computing group of the Cambridge University Engineering Department for their assistance with this work.

Cambridge, October 1974 R. K. LIVESLEY

†A list of references will be found on p. 270.

CHAPTER 1

Introduction

The concepts and notation of matrix algebra have for a long time been standard analytical tools of the applied mathematician. In the period before 1940 a few papers appeared in which these ideas were applied to structural problems, but in an age without automatic computers the approach attracted little attention from practising engineers. Indeed, a generation of designers which had recently been liberated from tedious manual calculations by the introduction of moment distribution was hardly likely to be enthusiastic about a method which required the formal manipulation of large arrays of coefficients.

The advent of the digital computer in the late 1940s produced a change in the criteria for judging whether a method of analysis was "good" or "bad". The question gradually ceased to be, "Does the method minimize the amount of numerical work?", and became, "Is the method one which can easily be organized for a computer?" The fact that methods based on matrix algebra are ideal in this respect accounts for the steady growth in their popularity over the last 20 years.

1.1. The value of a systematic approach

Structural theorists always have two aims. The first is to assist practical engineers, who have to design and build real structures in such a way that they are safe and economical, and who need rapid, simple and self-checking procedures for predicting stresses and deflections. The second is the traditional aim of the mathematician, intent on developing a consistent and logical framework of theory which will provide a general understanding of how structures behave. To a mathematician "understanding" a particular physical phenomenon means seeing it as an example of a more general mathematical concept—a concept which may also be the key to an apparently quite dissimilar physical problem. While such an approach may appear to have little to do with the design of efficient numerical procedures for practical use, it has an important function, for it is largely in this way that techniques developed in one engineering discipline are transferred to others.

In the past these two aims have sometimes seemed to be in conflict. The classical methods of structural analysis developed by men such as Maxwell, Mohr and Müller-Breslau certainly had the qualities of generality, logical simplicity and mathematical elegance. Unfortunately they often led to tedious calculations when applied to the analysis of practical structures, and in an age when even a desk calculating machine was a rarity this was a serious defect. Succeeding generations of engineers accordingly devoted a great deal of effort to simplifying this aspect of analysis. While many ingenious techniques of great practical value appeared, they were mostly specific to particular types of structure and inevitably the increasing number of superficially different methods tended to obscure the fundamental ideas from which the methods were all originally derived.

The introduction of the digital computer helped to resolve this conflict. The main objection to the earlier classical methods of analysis is that they lead to large systems of linear equations, which are difficult and tedious to solve by hand. With computers to do the numerical work this objection no longer has the same force. The advantages of a general approach, on the other hand, are accentuated, since a computer program is an expensive investment which can only be justified economically if it can be used repeatedly to analyse a large number of different structures.

Matrix methods provide the theoretician with a suitable framework for a general understanding of structural behaviour. They provide the practising engineer with systematic procedures of analysis which are easily converted into computer programs. One advantage of these procedures stems from the fact that virtually all computers have library routines for carrying out the standard operations of matrix algebra—multiplication, inversion, solution of equations, etc. From the programmer's point of view this means that it is almost as easy to program the multiplication of two matrices as to program the multiplication of two scalar numbers—the programmer need not even know the details of how the operation is carried out. In most places in this book we take the attitude that a problem is "solved" if it can be reduced to a series of standard matrix operations.

1.2. A structure as an assembly of elements

A complex engineering system is often analysed by regarding it as an assembly of elements, the properties of the system being determined from the relatively simple properties of the individual parts. In many problems the elements are identifiable pieces of engineering hardware—beams and columns in a steel framework, capacitors and transistors in an electronic circuit. In other cases where the physical system is continuous in nature the elements may be formed by the introduction of arbitrary boundaries in the material.

In either case an element has both "internal" and "external" characteristics. The internal characteristics are concerned with what really goes on inside the element. The external characteristics are concerned with what is observed at the boundaries, where the element interacts with its neighbours. It is the latter characteristics which determine the influence of the element on the overall behaviour of the system of which it is a part.

The elements used in structural analysis may be classified as

(a) line elements—beams, columns, arches, etc.;
(b) surface elements—plates, shells, regions of plane stress;
(c) volume elements—regions of general three-dimensional stress.

Whatever the type of structure, the process of dividing it into elements is to some extent under the control of the analyst. He may regard a structure as a large number of small elements or a small number of larger ones. If small elements are chosen then the number of variables in the analysis will be large, but a relatively crude approximate theory may suffice to predict the external characteristics of each individual element. If large elements are chosen then the number of variables needed to describe the overall behaviour of the structure will be less, but the characteristics of the elements will need to be represented more exactly. For example, if we wish to find the two-dimensional stress distribution in a gravity dam we may consider it as an assembly of triangular elements, as shown in Fig. 1.1. It is plausible to suggest that for the arrangement of elements in (a) we could obtain a satisfactory approximate solution by assuming constant stress within each element. However, to achieve a comparable degree of accuracy in the overall stress distribution using arrangement (b) would clearly require a consideration of the stress variation within each element.

This way of looking at structures, which we term the *finite element* approach, divides structural analysis into two kinds of activity. The first is associated with the properties of individual elements, and involves the translation of internal relationships between stresses and strains into external relationships between boundary forces and boundary displacements. The second is concerned with using these external relationships for individual elements to build up equations describing the behaviour of a complete structure. (We regard the solution of these equations simply as a mechanical process to be carried out by a computer.) Chapters 3 and 4 of this book are concerned with the properties of individual elements, while the remaining chapters are mainly concerned with complete structures.

1.3. Boundaries and nodes

When a number of finite elements are assembled to form a structure the physical process of joining the elements together corresponds to the imposition of conditions of displacement compatibility and stress continuity on the associated boundary variables. An essential part of the finite element approach is the replacement of these conditions by conditions at isolated points in the structure termed *nodes*.†

As an example of this process we consider the beam/column junction shown in Fig. 1.2. An exact analysis of this junction involves finding distributions of stress and displacement which satisfy appropriate continuity and compatibility conditions at every point of the boundary surface *AB*, while at the same time also satisfying the differential equations of elasticity within the beam and the column. Such an analysis is extremely

† In the sections of this book which apply to both continuum problems and skeletal structures we use the terms *elements* and *nodes*. When dealing specifically with skeletal structures we revert to the more familiar *members* and *joints*.

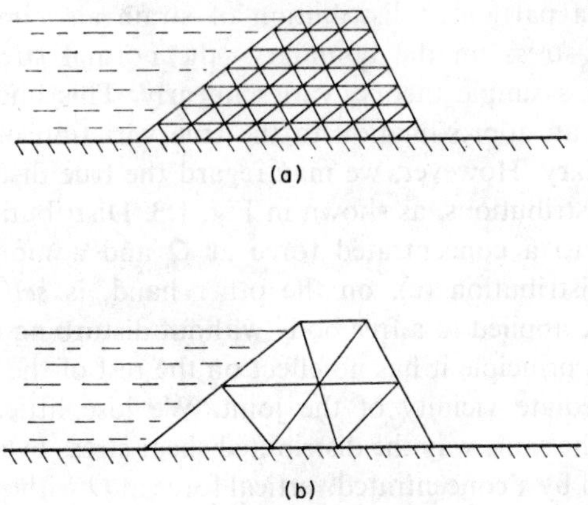

FIG. 1.1. Finite element representations of a continuum problem. (a) Fine mesh. (b) Coarse mesh.

complex. In practice, of course, we avoid the difficulties by making assumptions about the state of strain in the beam—we ignore transverse strains and adopt the engineer's familiar approximation that "plane sections remain plane".

These assumptions imply that the displacement of the boundary between the beam and the column is completely defined by the translation of the point *O* and the rotation of the plane surface *AB*—that is, by three parameters if the problem is regarded as a two-dimensional one. The

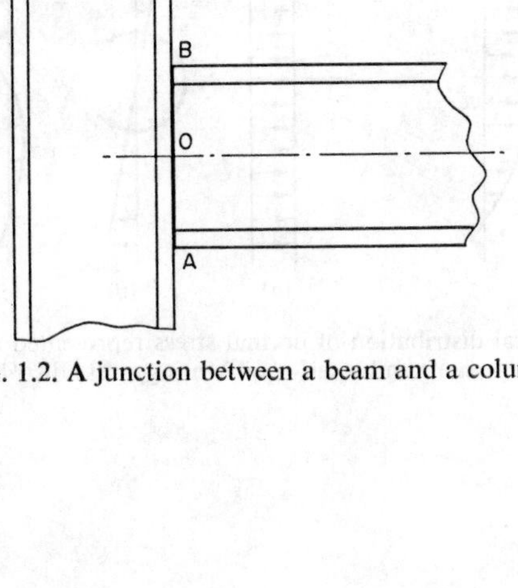

FIG. 1.2. A junction between a beam and a column.

assumption of a particular distribution of strain also implies a certain distribution of stress on the boundary—the normal stress component, according to this simple theory, varies linearly. This linear variation is obviously only an approximation to the true variation of normal stress over the boundary. However, we may regard the true distribution as the sum of three distributions, as shown in Fig. 1.3. Distributions (a) and (b) are equivalent to a concentrated force at O and a moment about O, respectively. Distribution (c), on the other hand, is *self-equilibrating*—that is, it can be applied to a free body without disturbing its equilibrium. By St. Venant's principle it has no effect on the rest of the structure away from the immediate vicinity of the joint. We lose little, therefore, by ignoring it. In the same way the distributed shear stress over the boundary may be replaced by a concentrated vertical force at O without affecting the behaviour of the structure away from the vicinity of the joint.

The end-result of the approximations we have made is that the compatibility conditions associated with the boundary AB are replaced by conditions on the two displacement components of the point O and a rotation about O. The stresses which act across the boundary are similarly replaced by two forces and a moment, which can be thought of as being concentrated at O—the familiar axial force, shear force and bending moment of simple bending theory, and it is these three quantities which appear in the equilibrium equations for the joint.

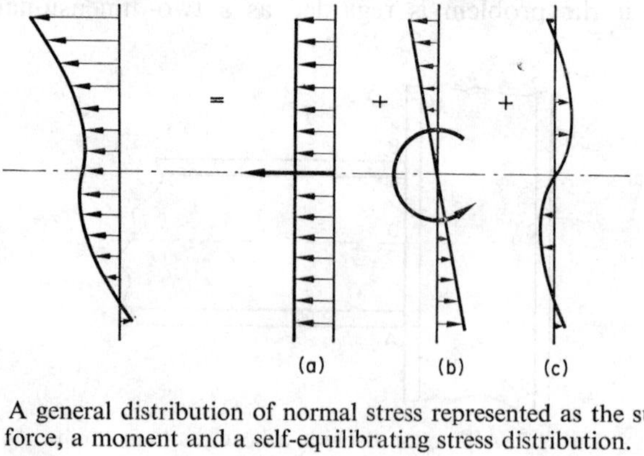

FIG. 1.3. A general distribution of normal stress represented as the sum of a force, a moment and a self-equilibrating stress distribution.

A similar procedure is used in the finite element treatment of a continuum problem. In Fig. 1.4 a continuum is represented by a series of plane triangular elements. Once again, an exact solution requires compatibility of displacements and continuity of stresses at every point on the boundary *AB*. To simplify the analysis we replace the compatibility condition by the condition that the components of displacement must be equal at the nodes *A* and *B*. At the same time we replace the stress continuity conditions by overall conditions of equilibrium which must be satisfied by concentrated forces at *A* and *B*, these forces being statically equivalent to the stresses on the boundaries which meet at those nodes.

It is apparent that these conditions allow discontinuities of both stress and displacement along a boundary. Discontinuities of stress are usually unimportant, since equilibrium is at least satisfied in an overall sense. (The out-of-balance stress distributions have only local effects, in the same way as distribution (c) in Fig. 1.3.) Discontinuities of displacement such as that shown in Fig. 1.4a are more serious. However, they may be avoided by restricting the mode of deformation of the elements in such a way that compatibility at nodes implies compatibility on intervening boundaries. For example, if in a plane triangular element we assume that stress and strain are constant throughout the element then all displacements vary linearly with position, so that all straight lines, including boundaries, remain straight. It follows that compatibility of displacement at two neighbouring nodes implies compatibility at all points on the boundary joining them, as shown in Fig. 1.4b. Such elements are said to be

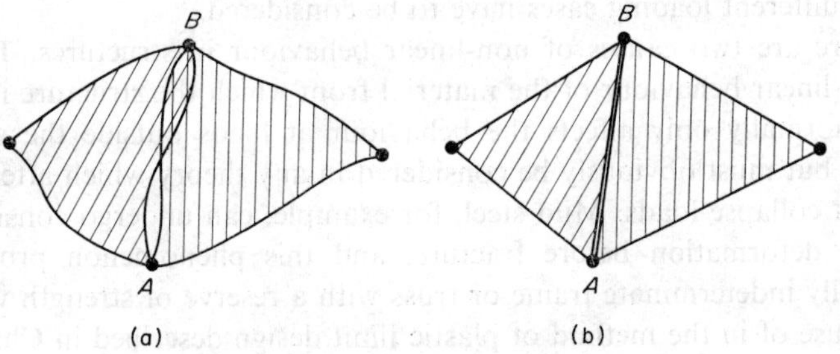

FIG. 1.4. A boundary between two plane stress elements. (a) Non-conforming elements. (b) Conforming elements.

conformable. Although non-conforming elements are sometimes used in practice, the discussion of area and volume elements in Chapter 4 is restricted to elements which are conformable.

The approximating procedures outlined in this section make the joints or nodes of a structure the focus of attention. The properties of the individual elements are described by relations between nodal loads and nodal displacements, the nodal loads being statically equivalent to the distributed boundary stresses which occur in the real structure. External loads acting on the structure are replaced by statically equivalent nodal loads, while compatibility and stress-continuity conditions along boundaries are replaced by compatibility and equilibrium conditions at nodes. The analysis of a structure is thus reduced to a problem with a finite number of degrees of freedom, for which the methods of matrix algebra are appropriate.

1.4. Linearity and superposition

An element or a structure is *linear* if all displacements and internal forces vary linearly with applied loads. Most real structures behave in an approximately linear manner under working loads, so that methods of analysis which assume linearity are important from a design engineer's viewpoint. The assumption of linearity has two important advantages. In the first place it makes the actual job of analysing a structure under a particular loading system a great deal easier. In the second place it allows the superposition of solutions, with a consequent saving of effort when many different loading cases have to be considered.

There are two causes of non-linear behaviour in structures. The first is non-linear behaviour of the material from which the structure is made. This normally only affects the behaviour at loads outside the working range, but must obviously be considered in any theory which attempts to predict collapse loads. Mild steel, for example, can undergo considerable plastic deformation before fracture, and this phenomenon provides a statically indeterminate frame or truss with a reserve of strength which is made use of in the method of plastic limit design described in Chapter 7.

The second cause is associated with changes of geometry. In linear analysis we always assume that the deformations of an element or a structure are "small". To be more precise, we assume that it is legitimate to write down all conditions of equilibrium in terms of the lengths and angles associated with the *undistorted* structure, whereas strictly these equations must hold in the *distorted* structure. Since the distortions are functions of the loading their inclusion in the equilibrium equations makes these equations non-linear. Non-linear behaviour due to changing geometry is often due to gross changes in the overall shape of the assembled structure. However, certain types of structural element can also behave in a non-linear manner, even when they are made of linear-elastic material. In a flexible cable, for example, the longitudinal stiffness varies with the tension, while the bending stiffness of a beam is a function of the axial force which it carries.

The most complex structural problems are those in which the various causes of non-linearity interact. In Chapter 10 we discuss a number of non-linear problems, and describe general techniques suitable for their analysis. In the remainder of the book, however, we shall assume linear-elastic material (except in Chapter 7) and ignore the affects of deformation when writing down equilibrium equations.

A linear theory of structures allows us to use the principle of superposition. This states that the stresses and deformations produced in a structure by a number of loads acting together can be obtained by adding up the stresses and deformations produced by each load acting separately. Thus we may determine the behaviour of a structure under a series of unit loads applied at different points and then calculate the effects of more complex loading patterns by combining these basic solutions. This procedure is not legitimate in the case of non-linear structures.

An important application of this principle occurs in the treatment of loads applied at internal points of structural elements. As stated in the previous section, the finite element approach focuses attention on the joints in a framework or the nodes in a continuum problem. It is at these points that compatibility and equilibrium conditions are applied, with loads applied at internal points of elements replaced by "equivalent" loads at the nodes. These equivalent loads may be determined as follows.

We take as an example the portal frame shown in Fig. 1.5. We replace the loading on the beam AB by the sum of two loadings, as shown in the figure. Loading (a) consists of the actual loads applied to the beam,

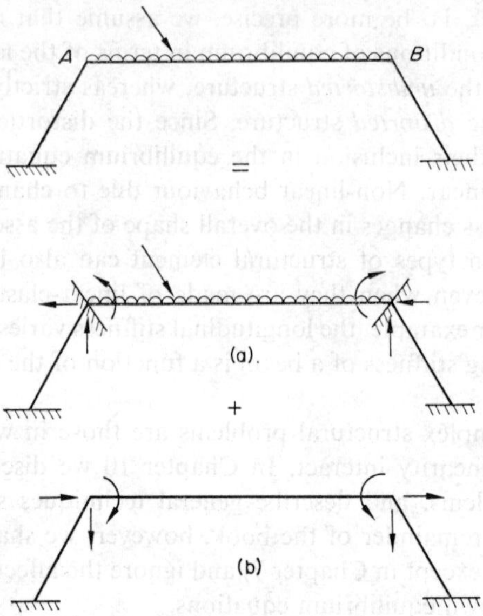

FIG. 1.5. A load system replaced by equivalent nodal loads.

together with a set of concentrated forces and moments acting at the joints *A* and *B*, these forces and moments being of such magnitudes as will prevent any translation or rotation of the two joints. Loading (b) consists of the concentrated forces and moments introduced in (a), reversed in sign. These forces and moments are termed the *equivalent nodal loads.*

Since in load system (a) neither of the joints is allowed to displace or rotate, the determination of the forces and moments applied at *A* and *B* may be carried out by treating the beam in isolation from the rest of the structure. Once the equivalent nodal loads have been found, the analysis of the structure under load system (b) can be carried out entirely in terms of the nodal variables. Finally the stresses and deformations associated with the two systems of loading are added to give the solution for the original loading.

This procedure may be carried out for any structure. As in the example, the calculation of the equivalent nodal loads which make up loading (a) may be carried out by considering each member individually, without reference to the rest of the structure. General formulae for these loads are given in Sections 3.10 and 4.4. It should be remembered that the procedure described above is dependent for its validity on the principle of superposition, which is only valid for linear structures.

A similar procedure may be applied to distributed mass and inertia in structural vibration problems. Once again the distributed inertia loading is replaced by equivalent inertia forces and moments at the nodes of the structure, so that subsequent analysis can be carried out entirely in terms of nodal quantities. Techniques for calculating these equivalent inertia loads are presented in Sections 3.11 and 4.6.

1.5. How analytical methods are built up

We have seen that in the finite element approach to structural analysis the properties of individual elements are represented by relationships between nodal loads and nodal displacements, and that external loads are replaced by equivalent concentrated loads at the nodes. By the *analysis* of a structure we normally mean the determination of the nodal loads and nodal displacements of all the elements for a given equivalent loading. (Note that the terms loads and displacements are used in a general sense here, implying moment and rotation components, where appropriate, as well as forces and translations.)

There are three types of equation which the nodal loads and nodal displacements of the elements must satisfy. These are

(a) The equations between nodal loads and nodal displacements for the individual elements. These are derived from the stress–strain equations of the element material.
(b) The nodal compatibility equations. These equate the appropriate nodal displacements of those elements which have a common node.
(c) The nodal equilibrium equations. These state that at each node the external load must equal the sum of the nodal loads acting on the elements which meet there.

To these we may add the subsidiary condition that for each element the nodal loads acting on it must be such as to keep it in overall equilibrium. As we shall see in Chapters 3 and 4, this condition is usually satisfied automatically during the development of the element load/displacement equations. If the equilibrium conditions (c) provide sufficient equations to define all the internal loads uniquely then a structure is said to be *statically determinate*. If not, it is said to be *statically indeterminate* or *hyperstatic*.

The analysis of a determinate structure is relatively simple, since all the internal loads may be determined from equilibrium considerations alone. Conditions (a) and (b) may be applied afterwards if the displacements of the structure are required. In analysing a hyperstatic structure, on the other hand, it is necessary to use all three conditions in order to obtain either stresses or displacements. Broadly speaking, one may classify methods of structural analysis according to the *order* in which the conditions of equilibrium and compatibility are applied. Methods in which the compatibility conditions are used first give rise to equations of joint equilibrium, and are called *equilibrium* or *displacement* methods. Methods in which the equilibrium conditions are satisfied first lead to equations of displacement compatibility and are called *compatibility* or *force* methods.

The essence of the equilibrium approach is that the nodal displacements are regarded as the basic unknowns. The compatibility equations (b) are combined with the element load/displacement equations (a) to give expressions for the nodal loads on the elements in terms of the nodal displacements. These expressions are then substituted into the equilibrium equations (c). The result is a set of equations relating the external loads (i.e. the concentrated nodal loads equivalent to the external loading) to the nodal displacements. These equations are solved for the nodal displacements, after which the internal nodal loads can be found from the individual element equations (a).

It is apparent that in an application of the equilibrium method it is always the displacements which are computed first. The number of equations which have to be solved is equal to the total number of independent displacement variables—the *number of degrees of freedom* of the structure, as it is often called, and is not affected by whether the structure is statically determinate or hyperstatic. This method is described in Chapter 5.

In the compatibility approach, on the other hand, use is made of the fact that it is relatively easy to analyse a determinate structure. Even when a structure is hyperstatic the equations of nodal equilibrium still hold, and these equations may be used to express the complete set of element nodal loads in terms of a smaller number of unknowns, often termed the *redundant* forces and moments.

While the equilibrium approach uses the load/displacement equations for the elements to express nodal loads in terms of nodal displacements, the compatibility approach uses these relationships in an inverse form, deformations being expressed in terms of nodal loads. These loads are then expressed in terms of the redundant forces and moments and the known external loads by means of the conditions of equilibrium. Thus we obtain all the member deformations as functions of the external loads and the unknown redundants.

The next step involves the application of the compatibility conditions to these deformations. The result is a set of equations which in effect states that the structure is continuous at the points at which the redundant forces and moments act. Solution of these equations gives the values of the redundant forces and moments, and hence all the nodal loads. The expressions for the element deformations in terms of the applied loads and the redundants may then be used to find the displacements of the nodes. This method is described in Chapter 8.

In comparing the equilibrium and compatibility approaches the most obvious test is to consider the number of equations which have to be solved. In an equilibrium analysis of a structure the number of equations is equal to the number of degrees of freedom, while in a compatibility analysis it is equal to the number of redundants. However, a comparison on these grounds may well be misleading, since the ease with which large systems of equations can be solved on a modern computer makes the actual solution process a relatively trivial part of the complete analysis. It is more sensible to base a comparison on the amount of work which has to be done in both setting up and solving the equations, and the ease with which this work can be systematized. A comparison is also affected by the arrangement of coefficients in the various matrices which have to be handled. As we shall see in Section 11.1, the equilibrium method tends to produce matrices with a large number of zero coefficients, with the non-zero terms grouped in bands close to the leading diagonal. It is possible to develop computing procedures which take advantage of this feature, enabling large problems to be solved on relatively small computers. Although a compatibility approach often involves fewer variables the matrices tend to be more densely populated.

Although the equilibrium and compatibility methods are perhaps the most clearly defined approaches to structural analysis, they are best regarded as the extreme ends of a spectrum of methods in which the three types of equation are combined in a variety of ways. A typical example of an intermediate approach is the method of transfer matrices described in Chapter 9.

CHAPTER 2

The Main Variables and Relationships

In this chapter we introduce the variables which form the basis of much of the later analysis, and discuss the most important relationships between them. Although most of the detailed treatment in later chapters is concerned with "line elements", i.e. beams and columns, at this stage our aim is to establish a notation suitable for discussing all types of element. A summary of the notation will be found on p. 238.

Throughout the book, vectors associated with individual elements or individual nodes are set in italic type, while vectors relating to complete structures are set in roman type. When considering a general structure we label the joints or nodes which can displace under the action of applied loads with capital letters *A*,*B*, . . . (or numbers 1,2, . .). For the moment we regard the actual assignment of letters or numbers as being done in an arbitrary manner, although there are computational advantages in adopting certain patterns of numbering (see Section 11.1). Points at which the structure is attached to a rigid foundation are labelled *O* (or 0). In the same way we identify the elements by lower case letters *a*,*b*, ... When considering the properties of individual elements we number the boundary nodes 1,2, ...

2.1. Nodal variables—loads and displacements

As mentioned in Section 1.3, the finite element approach assumes that an element of a structure is related to the rest of the structure entirely through the nodal loads and displacements. The relationship between these variables provides us with an "external" view of the properties of the element.

Throughout the rest of this book we shall use the term *displacement* in an extended sense, implying rotation, where appropriate, as well as translation. Thus in a rigid-jointed space framework the displacement of a node is a vector of six components—three translations and three rotations—while in a plane pin-jointed truss or plane finite-element assembly it merely consists of the two components of translation. We shall use the symbol $\boldsymbol{d}$ when referring to displacement vectors associated with individual nodes or structural elements, with appropriate suffices where necessary. When a number of such vectors are combined to form a vector representing the displacement of a complete structure we shall represent this by the symbol $\mathbf{d}$.

In the same way the term *load* will, where appropriate, imply moments as well as forces. For each component in a displacement vector there will be a component in the corresponding load vector, forces being associated with translations and moments with rotations. We shall use the symbol $\boldsymbol{p}$ when referring to an individual node or element and the symbol $\mathbf{p}$ when referring to the complete set of loads on a structure. In diagrams single arrows will be used to represent applied loads and nodal displacements, the existence of moments and rotations being implied where appropriate.

Throughout the book we shall use right-handed coordinate axes x, y, z. Thus in a space frame with rigid joints the displacement and load vectors will have the form

$$\boldsymbol{d} = \begin{bmatrix} \delta_x \\ \delta_y \\ \delta_z \\ \theta_x \\ \theta_y \\ \theta_z \end{bmatrix}, \qquad \boldsymbol{p} = \begin{bmatrix} p_x \\ p_y \\ p_z \\ m_x \\ m_y \\ m_z \end{bmatrix}$$

where δ, θ, p and m indicate translation, rotation, force and moment components respectively, and the sign convention is as shown in Fig. 2.1. In a plane framework the corresponding coordinate system is that shown in Fig. 2.2, the displacement and load vectors being

$$\boldsymbol{d} = \begin{bmatrix} \delta_x \\ \delta_y \\ \theta \end{bmatrix}, \qquad \boldsymbol{p} = \begin{bmatrix} p_x \\ p_y \\ m \end{bmatrix}$$

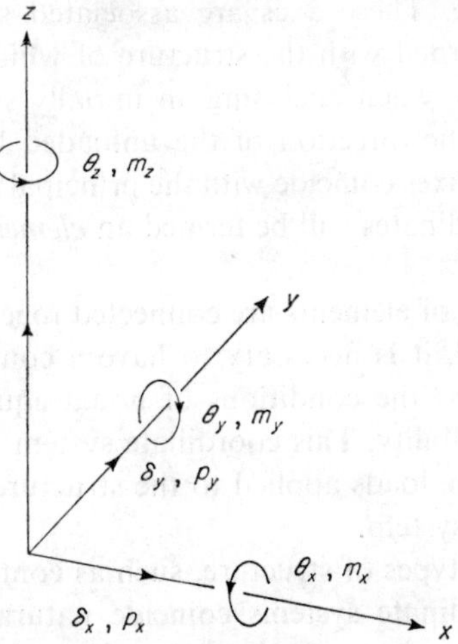

FIG. 2.1. Coordinate system for a three-dimensional structure, showing sign conventions for displacement and load components.

As we shall see later, the principle of virtual work provides a powerful tool for deriving general relationships. For this tool to function correctly it is necessary for the components of loads and displacements to correspond in a work sense. Thus the units and sign conventions associated with these components must be such that a load $\boldsymbol{p}$ moving through a displacement $\boldsymbol{d}$ does work $\boldsymbol{p}^t\boldsymbol{d}$. If this correspondence is maintained then the relationships we shall derive will be true in any consistent set of units.

When deriving load/displacement equations for an element it is natural to choose coordinate axes which make the form of the equations as

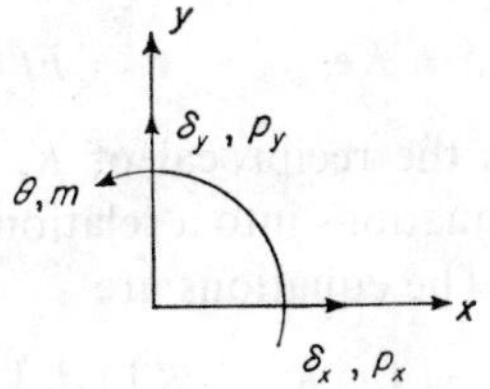

FIG. 2.2. Coordinate system for a plane structure.

simple as possible. These axes are associated solely with the element, and are not concerned with the structure of which the element may be a part. For example, when analysing an initially straight beam one would naturally choose the direction of the unloaded beam as the x-axis, and make the y- and z-axes coincide with the principal axes of the cross-section. Such a set of coordinates will be termed an *element* or *member* coordinate system.

When a number of elements are connected together to form a structure, on the other hand, it is necessary to have a common coordinate system in which to express the conditions of nodal equilibrium and nodal displacement compatibility. This coordinate system is the natural one to use when specifying the loads applied to the structure, and is referred to as a *global* coordinate system.

There are some types of structure, such as continuous beams, where all the member coordinate systems coincide naturally with the global coordinate system. In most cases, however, the various systems are different, and it is necessary, at some stage in the analysis, to change from element to global coordinates. One advantage of matrix notation is the ease with which transformations of this sort may be made. The matrices commonly associated with coordinate transformations are introduced in Section 2.4.

2.2. Element variables—stress-resultants and deformations

The concepts of "stiffness" and "flexibility" are introduced in elementary mechanics through the study of a linear one-dimensional elastic bar, such as the one shown in Fig. 2.3. We write

tension = stiffness × extension, extension = flexibility × tension

or

$$t = Ke, \qquad e = Ft \tag{2.1a, 2.1b}$$

where F, the flexibility, is the reciprocal of K, the stiffness. It is easy to convert the first of these equations into a relationship between nodal loads and nodal displacements. The equations are

$$\begin{bmatrix} p_1 \\ p_2 \end{bmatrix} = \begin{bmatrix} K & -K \\ -K & K \end{bmatrix} \begin{bmatrix} d_1 \\ d_2 \end{bmatrix} \tag{2.2}$$

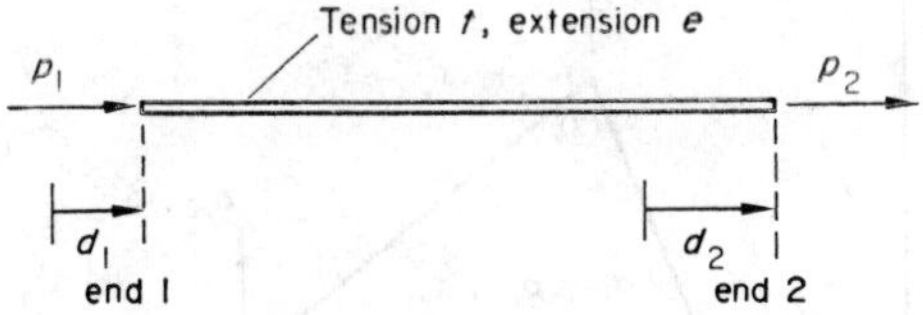

FIG. 2.3. The variables associated with an elastic bar in one dimension.

We shall see in Chapter 5 that this is the form required by the "equilibrium" or "displacement" method of analysis. We may think of (2.1a) and (2.1b) as providing an "internal" view of the properties of the element, in contrast to the "external" view provided by the relationship between the loads and displacements in (2.2). The matrix appearing in (2.2) is referred to as the *nodal stiffness matrix* of the bar because it relates nodal loads to nodal displacements. However, it would be stupid to regard this matrix as providing the fundamental definition of the properties of the bar, since all the essential information is contained in the single number K. It is not simply that the matrix repeats the same information four times. Equally important, the matrix

$$\begin{bmatrix} K & -K \\ -K & K \end{bmatrix}$$

is singular and therefore has no inverse. This is because p_1 and p_2 are not linearly independent, since they must satisfy the equilibrium condition $p_1 = -p_2$, whatever the value of K. Alternatively we may say that the bar may be given an arbitrary rigid-body displacement without affecting the values of p_1 and p_2, so that we cannot solve equation (2.2) for d_1 and d_2.

The same pattern emerges whatever the element. For example, consider the plane triangular element shown in Fig. 2.4, in which the three components of stress σ_x, σ_y, τ_{xy} and the three components of strain ϵ_x, ϵ_y, γ_{xy} are assumed to be constant throughout the element. The stresses are statically equivalent to the nodal forces p_{x1}, p_{y1}, . . ., while the strains are compatible with the nodal translations δ_{x1}, δ_{y1}, ... The "internal"

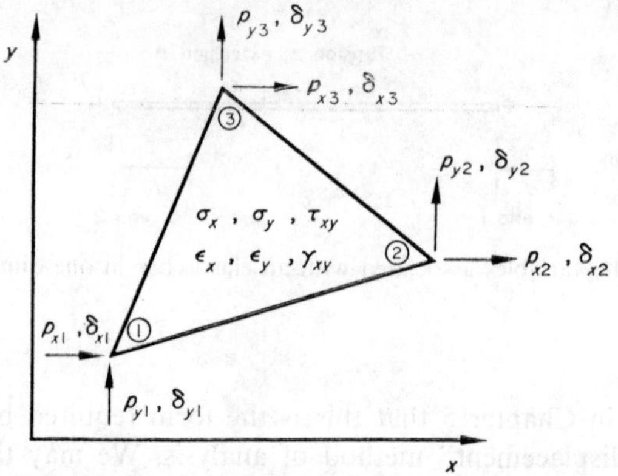

FIG. 2.4. The variables associated with a constant-strain plane triangular element.

description of the elastic properties of the element is given by the equations†

$$\begin{bmatrix} \sigma_x \\ \sigma_y \\ \tau_{xy} \end{bmatrix} = \frac{E}{1-\nu^2}\begin{bmatrix} 1 & \nu & 0 \\ \nu & 1 & 0 \\ 0 & 0 & \dfrac{1-\nu}{2} \end{bmatrix}\begin{bmatrix} \epsilon_x \\ \epsilon_y \\ \gamma_{xy} \end{bmatrix} \qquad (2.3)$$

The 3×3 matrix in these equations (or its inverse) defines the elastic properties of the material, while the components of the stress and strain vectors define the particular state of the element. In contrast, the "external" description of the properties of the element is given by the equations

$$\begin{bmatrix} p_{x1} \\ p_{y1} \\ p_{x2} \\ p_{y2} \\ p_{x3} \\ p_{y3} \end{bmatrix} = \begin{bmatrix} \\ \\ 6 \times 6 \\ \text{Matrix} \\ \\ \\ \end{bmatrix}\begin{bmatrix} \delta_{x1} \\ \delta_{y1} \\ \delta_{x2} \\ \delta_{y2} \\ \delta_{x3} \\ \delta_{y3} \end{bmatrix} \qquad (2.4)$$

† The reader may be somewhat surprised to see stresses and strains written as vectors. This is legitimate provided that we realize that such quantities are not "geometrical vectors" in the sense of obeying the vector rules associated with the geometrical space occupied by the element. Stress, for example, is certainly not "a quantity possessing magnitude and direction" in the sense that force is. This distinction is important when we make a change from one geometrical coordinate system to another. In practice we only carry out such transformations when we have reached the stage in analysis at which we are dealing with geometrical vectors, i.e. loads and displacements.

Clearly the nodal stiffness matrix in (2.4) does not really contain any more information about the material properties than the 3 × 3 matrix in (2.3). Equations (2.4), like (2.2), cannot be solved for the displacements, since the element may be given a rigid-body displacement without altering the nodal loads. In fact only three of the p's are linearly independent, since the six components of the nodal load vector $\boldsymbol{p}$ must satisfy the three conditions of element equilibrium.

Whatever the structural element, it is clearly desirable to define its stiffness properties as concisely as possible. The fact that the components of the nodal load vector $\boldsymbol{p}$ are not linearly independent means that it is possible to express $\boldsymbol{p}$ in terms of a vector of fewer components. Generalizing from our two examples, we introduce the concept of the *element stress-resultant vector* $\boldsymbol{r}$.

This quantity is simply a vector, whose components are linearly independent, which defines the state of stress throughout the element. Its components may be stresses, stress-resultants (i.e. integrals of stresses over an area) or simply algebraic parameters in some general stress function. Simple examples are t in equation (2.1) and the vector of stresses in equation (2.3). For a given element the nodal load vector $\boldsymbol{p}$ can always be expressed in terms of $\boldsymbol{r}$ from the equilibrium equations of the element. Since the vector $\boldsymbol{p}$ must satisfy these equations, the number of components in $\boldsymbol{p}$ will always be greater than the number in $\boldsymbol{r}$—one greater for a simple bar, three greater for an element in a plane, six greater for a general element in space. There is considerable freedom of choice in selecting the variables which make up $\boldsymbol{r}$ in any given element. Thus for the simple triangle of Fig. 2.4 we could choose the stress vector appearing in equation (2.3), or some alternative representation of it (i.e. one, say, in which σ_x and σ_y were replaced by the hydrostatic and deviatoric stresses). Alternatively we could choose three of the six nodal load components. (Note, however, that the choice is not entirely arbitrary. The components p_{x1}, p_{x2}, p_{x3}, for example, are not linearly independent.)

We now define a vector $\boldsymbol{e}$, with the same number of components as $\boldsymbol{r}$, which we call the *element deformation vector*. This vector specifies the deformation of the element uniquely. It must "correspond" to $\boldsymbol{r}$, in the sense that in a virtual deformation $\boldsymbol{e}^*$ the virtual work done within the element must be $\boldsymbol{r}^t\boldsymbol{e}^*$. This requirement that $\boldsymbol{r}$ and $\boldsymbol{e}$ must correspond in a work sense means that a choice of components for one vector defines by implication the components of the other. The importance of the deformation vector lies in the fact that it is not affected by any rigid-body displacement which the element may have. A simple example is the variable e in equation (2.1). Note that the stress and strain vectors in (2.3) are not a suitable pair as they stand, since they do not satisfy the above condition. To obtain proper correspondence one or other must be multiplied by the volume of the element.

Having defined the vectors $\boldsymbol{r}$ and $\boldsymbol{e}$, we may specify the elastic properties of a general element by the equations

$$\boldsymbol{r} = \boldsymbol{K}\boldsymbol{e}, \qquad \boldsymbol{e} = \boldsymbol{F}\boldsymbol{r} \tag{2.5}$$

where $\boldsymbol{F} = \boldsymbol{K}^{-1}$. $\boldsymbol{K}$ and $\boldsymbol{F}$ are always symmetric by virtue of the reciprocal theorem, and may be made diagonal by a suitable choice of components for $\boldsymbol{r}$ and $\boldsymbol{e}$. Details of $\boldsymbol{K}$ and $\boldsymbol{F}$ matrices for a number of elements are derived in Chapters 3 and 4.

The term "stress-resultant" simply means a stress integrated over an area or throughout a volume. It is important to realize that the terms "stress-resultant" and "load" are not synonymous. A load is something which acts on an element *from the outside*. Conventionally it is represented graphically by a directed arrow, and its components are specified as positive or negative in accordance with a sign convention such as that shown in Fig. 2.1. A stress-resultant, on the other hand, is essentially something *internal* to an element. A typical example is a pair of equal and opposite forces or moments, acting on the two sides of an imaginary cut in an element, as shown in Fig. 2.5. This pair of forces or moments is essentially self-equilibrating, and may be represented on a diagram by a double-ended arrow. Clearly the sign convention of Fig. 2.1 is inappro-

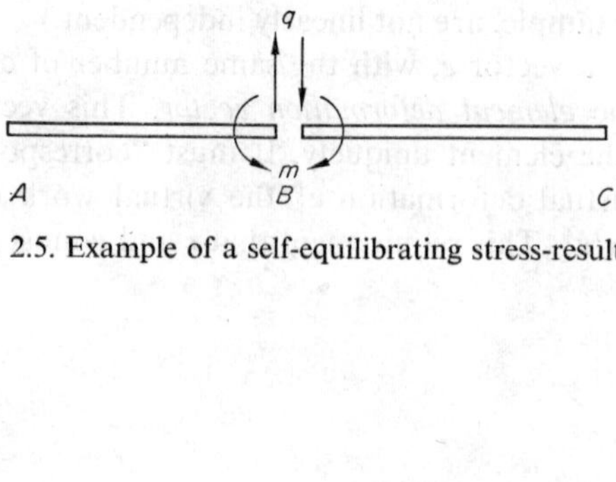

FIG. 2.5. Example of a self-equilibrating stress-resultant.

priate for such quantities. Instead we use conventions such as "tensile force is positive" and "sagging bending moment is positive".

Whether a quantity is a stress-resultant or a load may depend on how much of a structure we are considering. For example, if we consider the beam *AC* in Fig. 2.5 as a single element then m and q are typical "internal" self-equilibrating stress-resultants. However, if we consider *AB* and *BC* as separate elements then *AB* is acted on at *B* by an "external" load

$$\begin{bmatrix} q \\ m \end{bmatrix}$$

while *BC* is acted on by another "external" load

$$\begin{bmatrix} -q \\ -m \end{bmatrix}$$

2.3. Equilibrium and compatibility conditions: virtual work

In any analysis based on a finite element philosophy there are two types of equilibrium condition to be satisfied.

The first type is associated with elements. The nodal loads acting on each element must be such as will keep it in equilibrium. These conditions are satisfied automatically if we express the nodal loads in terms of the element stress-resultant vector. For example, in the bar of Fig. 2.3 the equilibrium condition $p_1 = -p_2$ is implied by the equation

$$\begin{bmatrix} p_1 \\ p_2 \end{bmatrix} = \begin{bmatrix} -1 \\ 1 \end{bmatrix} t \tag{2.6}$$

In a general element with n nodes the equilibrium conditions may be written in the form

$$\boldsymbol{p} = \boldsymbol{H}\boldsymbol{r} \tag{2.7}$$

or equivalently as

$$\left.\begin{aligned} \boldsymbol{p}_1 &= \boldsymbol{H}_1\boldsymbol{r} \\ \boldsymbol{p}_2 &= \boldsymbol{H}_2\boldsymbol{r} \\ \cdot\; &\quad \cdot\,\cdot \\ \cdot\; &\quad \cdot\,\cdot \\ \boldsymbol{p}_n &= \boldsymbol{H}_n\boldsymbol{r} \end{aligned}\right\} \tag{2.8}$$

where the suffices 1 . . . n indicate the individual nodes of the element. The matrices $\boldsymbol{H}_1$, $\boldsymbol{H}_2$, etc., are merely the appropriate submatrices (i.e. rows) of $\boldsymbol{H}$. We shall derive the detailed form of $\boldsymbol{H}$ for a number of different elements in Chapters 3 and 4.

The second type of equilibrium equation is associated with joints or nodes. At each node the external load must equal the sum of the nodal loads acting on the elements which meet there. If we consider the system of two bars shown in Fig. 2.6 the condition at the single node A may be written as $p_A = p_{2a} + p_{1b}$ or as

$$p_A = [1 \quad 1] \begin{bmatrix} p_{2a} \\ p_{1b} \end{bmatrix} \tag{2.9}$$

The two types of condition may be combined by writing the nodal loads appearing in (2.9) in terms of the appropriate element stress-resultant vectors (i.e. the tensions t_a, t_b). This gives

$$p_A = [1 \quad -1] \begin{bmatrix} t_a \\ t_b \end{bmatrix} \tag{2.10}$$

In a general structure we can set up similar relationships between the vector $\mathbf{p}$ consisting of all the applied nodal loads and the vector $\mathbf{r}$ formed from all the element stress-resultants. This may be written as

$$\mathbf{p} = \mathbf{H}\mathbf{r}. \tag{2.11}$$

We shall discuss general techniques for setting up the matrix $\mathbf{H}$ in Chapter 6.

Associated with each of these equilibrium equations there is a corresponding compatibility equation. We consider first the single element of Fig. 2.3, for which the equilibrium equation is (2.6). The compatibility

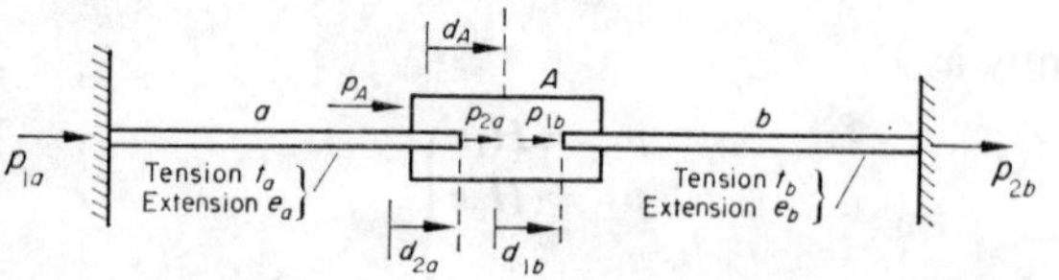

FIG. 2.6. A simple structure consisting of two bars a and b with a common node A.

equation relates the deformation e to the nodal displacements d_1, d_2 and is obviously $e = d_2 - d_1$. If we write this out in matrix form

$$e = [-1 \quad 1] \begin{bmatrix} d_1 \\ d_2 \end{bmatrix} \tag{2.12}$$

we see that the coefficient matrix which appears in (2.12) is the transpose of the coefficient matrix in (2.6).

Next we consider the condition for compatibility of displacement at the node A in Fig. 2.6. This is simply $d_{2a} = d_{1b} = d_A$, which may be written as

$$\begin{bmatrix} d_{2a} \\ d_{1b} \end{bmatrix} = \begin{bmatrix} 1 \\ 1 \end{bmatrix} d_A \tag{2.13}$$

and once again we find that the coefficient matrix in (2.9) reappears transposed in (2.13). Finally if we replace d_{2a} and d_{1b} in (2.13) by e_a and $-e_b$ we obtain

$$\begin{bmatrix} e_a \\ e_b \end{bmatrix} = \begin{bmatrix} 1 \\ -1 \end{bmatrix} d_A \tag{2.14}$$

which we compare with (2.10).

The duality which we observe in these three examples suggests that a similar duality between equilibrium and compatibility equations might exist in general. It is easy to prove that this is indeed the case. Consider, for instance, equation (2.7), which relates the nodal load vector $\boldsymbol{p}$ and the stress-resultant vector $\boldsymbol{r}$ for a general element. Imagine that the nodes are given virtual displacements $\boldsymbol{d}^*$, and that the associated element deformations are $\boldsymbol{e}^*$. Looking at the element "from the outside" the work done by the applied forces is $\boldsymbol{p}^t\boldsymbol{d}^*$. Looking "from the inside" the work done is $\boldsymbol{r}^t\boldsymbol{e}^*$ (remember that this is how we defined $\boldsymbol{e}$). Equating the two gives

$$\boldsymbol{p}^t\boldsymbol{d}^* = \boldsymbol{r}^t\boldsymbol{e}^*. \tag{2.15}$$

If we now write (2.7) in the form $\boldsymbol{p}^t = \boldsymbol{r}^t\boldsymbol{H}^t$ and substitute this in (2.15) we obtain

$$\boldsymbol{r}^t\boldsymbol{H}^t\boldsymbol{d}^* = \boldsymbol{r}^t\boldsymbol{e}^*.$$

Since the virtual displacement could be applied for any pair of values $\boldsymbol{p}$ and $\boldsymbol{r}$ it follows that

$$\boldsymbol{e}^* = \boldsymbol{H}^t\boldsymbol{d}^*. \tag{2.16a}$$

So far the argument has made no assumption about linearity. If now we make the usual assumption of displacements and strains being small we may replace the *virtual* quantities $\boldsymbol{e}^*$ and $\boldsymbol{d}^*$ by the corresponding *real* quantities and write

$$\boldsymbol{e} = \boldsymbol{H}^t\boldsymbol{d} \tag{2.16b}$$

which we compare with (2.7). We can regard (2.12) as a special case of (2.16b), just as (2.6) is a special case of (2.7). We may write (2.16b) in a form which corresponds to (2.8)

$$\boldsymbol{e} = \boldsymbol{H}_1{}^t\boldsymbol{d}_1 + \boldsymbol{H}_2{}^t\boldsymbol{d}_2 + \ldots + \boldsymbol{H}_n{}^t\boldsymbol{d}_n \tag{2.17}$$

where $\boldsymbol{d}_1, \ldots, \boldsymbol{d}_n$ are the displacements of the individual nodes of the element. A very similar argument starting from equation (2.11) shows that the set of nodal compatibility conditions for a complete structure may be written in the form

$$\mathbf{e} = \mathbf{H}^t\mathbf{d}. \tag{2.18}$$

Note that the step from (2.16a) to (2.16b) is equivalent to assuming that $\boldsymbol{H}$ is constant during a real deformation, i.e. that the equilibrium equations for the *undeformed* element may be used when the element is *deformed*. We make this assumption consistently in Chapters 3 to 9. Techniques which take account of the variation of $\boldsymbol{H}$ with deformation are discussed in Chapter 10.

This analysis is important because it shows that conditions of compatibility and equilibrium are not independent, but are very closely related. In later chapters we shall often use reasoning similar to that given above. The relationship between equilibrium and compatibility conditions typified by equations (2.7) and (2.16b) or (2.11) and (2.18) is known as *contragredience*.

2.4. Coordinate transformations

We now consider the problem of changing equations which have been developed in an element coordinate system into equations which relate vectors expressed in global coordinates.

Figure 2.7 shows a node i of a plane-stress finite element. In element coordinates x, y the load at the node is written as

$$\boldsymbol{p}_i = \begin{bmatrix} p_{xi} \\ p_{yi} \end{bmatrix}$$

while in global coordinates x', y' it becomes

$$\boldsymbol{p}_i' = \begin{bmatrix} p_{xi}' \\ p_{yi}' \end{bmatrix}$$

Resolving in the x' and y' directions we obtain

$$\begin{bmatrix} p_{xi}' \\ p_{yi}' \end{bmatrix} = \begin{bmatrix} \cos\alpha & -\sin\alpha \\ \sin\alpha & \cos\alpha \end{bmatrix} \begin{bmatrix} p_{xi} \\ p_{yi} \end{bmatrix} \tag{2.19a}$$

which we write as

$$\boldsymbol{p}_i' = \boldsymbol{T}\boldsymbol{p}_i. \tag{2.19b}$$

Since $\boldsymbol{p}_i$ is related to the element stress-resultant vector $\boldsymbol{r}$ by the equation $\boldsymbol{p}_i = \boldsymbol{H}_i\boldsymbol{r}$ we may write (2.19b) as

$$\boldsymbol{p}_i' = \boldsymbol{T}\boldsymbol{H}_i\boldsymbol{r} = \boldsymbol{H}_i'\boldsymbol{r} \tag{2.20}$$

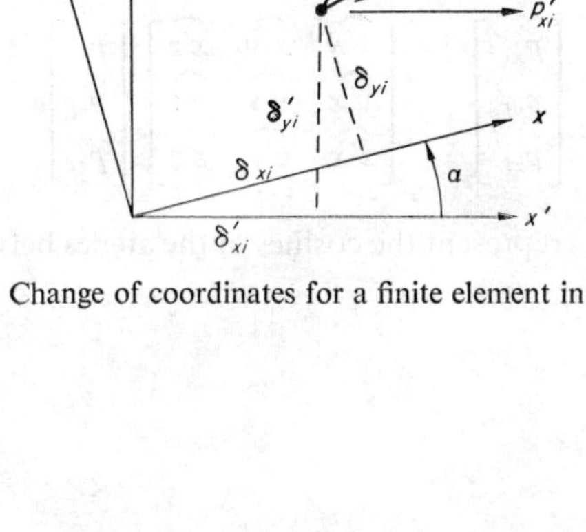

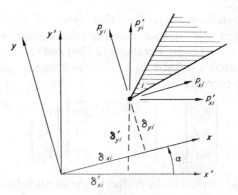

FIG. 2.7. Change of coordinates for a finite element in a plane.

where $\boldsymbol{H}_i' = \boldsymbol{T}\boldsymbol{H}_i$. Note that in (2.20) it is only $\boldsymbol{p}_i$ which has been transformed into the coordinate system x' y'. As mentioned in the footnote on p. 20, $\boldsymbol{r}$ is not in general a "geometrical" vector and cannot be transformed using the matrix $\boldsymbol{T}$.

The displacement components are similarly related by the equations

$$\begin{bmatrix} \delta_{xi} \\ \delta_{yi} \end{bmatrix} = \begin{bmatrix} \cos\alpha & \sin\alpha \\ -\sin\alpha & \cos\alpha \end{bmatrix} \begin{bmatrix} \delta_{xi}' \\ \delta_{yi}' \end{bmatrix} \tag{2.21a}$$

or

$$\boldsymbol{d}_i = \boldsymbol{T}^t\boldsymbol{d}_i'. \tag{2.21b}$$

Combining (2.21b) with (2.17) we obtain

$$\boldsymbol{e} = \boldsymbol{H}_1{}^t\boldsymbol{T}^t\boldsymbol{d}_1' + \boldsymbol{H}_2{}^t\boldsymbol{T}^t\boldsymbol{d}_2' + \ldots \boldsymbol{H}_n{}^t\boldsymbol{T}^t\boldsymbol{d}_n' \tag{2.22a}$$

$$= (\boldsymbol{H}_1')^t\boldsymbol{d}_1' + (\boldsymbol{H}_2')^t\boldsymbol{d}_2' + \ldots (\boldsymbol{H}_n')^t\boldsymbol{d}_n'. \tag{2.22b}$$

Equations (2.21) may be deduced directly from Fig. 2.7 or more generally by a virtual work argument of the type used previously in this chapter. Clearly the work done in a virtual displacement must be the same whatever coordinate system is used, so that

$$\boldsymbol{p}_i{}^t\boldsymbol{d}_i^* = (\boldsymbol{p}_i')^t\boldsymbol{d}_i'^* = \boldsymbol{p}_i{}^t\boldsymbol{T}^t\boldsymbol{d}_i'^*.$$

Hence $\boldsymbol{d}_i^* = \boldsymbol{T}^t\boldsymbol{d}_i'^*$, and replacing virtual displacements by real ones we have equation (2.21b).

Coordinate transformations for other types of element follow the pattern of this example. In each case the relationships between the two pairs of vectors are given by equations (2.19b) and (2.21b), provided that the matrix $\boldsymbol{T}$ is suitably defined. In a pin-jointed space frame, for example, equation (2.19a) becomes

$$\begin{bmatrix} p_{xi}' \\ p_{yi}' \\ p_{zi}' \end{bmatrix} = \begin{bmatrix} \widehat{x'x} & \widehat{x'y} & \widehat{x'z} \\ \widehat{y'x} & \widehat{y'y} & \widehat{y'z} \\ \widehat{z'x} & \widehat{z'y} & \widehat{z'z} \end{bmatrix} \begin{bmatrix} p_{xi} \\ p_{yi} \\ p_{zi} \end{bmatrix} \tag{2.23}$$

where $\widehat{x'x}$, $\widehat{x'y}$, etc., represent the cosines of the angles between the specified axes.

In a change from one system of rectangular coordinates to another the matrix $\boldsymbol{T}$ is nearly always orthogonal.† It follows that $\boldsymbol{T}^{-1} = \boldsymbol{T}^t$, and we may therefore write (2.19b) and (2.21b) in the inverse form

$$\boldsymbol{p}_i = \boldsymbol{T}^t\boldsymbol{p}_i', \qquad \boldsymbol{d}_i' = \boldsymbol{T}\boldsymbol{d}_i \qquad (2.24a), (2.24b)$$

† This is not true for the transformations associated with pin-ended bars. The relevant analysis for these elements is given in Section 3.4.

CHAPTER 3

The Elastic Properties of Single Elements

(a) *Line Elements*

In the previous chapter we introduced the idea of describing the elastic properties of a structural element by the equations

$$\boldsymbol{r} = \boldsymbol{K}\boldsymbol{e}, \qquad \boldsymbol{e} = \boldsymbol{F}\boldsymbol{r} \tag{3.1a, 3.1b}$$

where $\boldsymbol{r}$, the stress-resultant vector, defines the state of stress in the element and $\boldsymbol{e}$, the deformation vector, defines the corresponding deformations. We argued that these equations are the best way of defining the element properties because

1. the components of $\boldsymbol{r}$ are linearly independent;
2. the vector $\boldsymbol{e}$ is independent of any rigid-body displacement which the element may have;
3. the matrix $\boldsymbol{K}$ is always non-singular, so that the inverse matrix $\boldsymbol{F}$ always exists.

Also associated with the element are the equilibrium equations

$$\boldsymbol{p} = \boldsymbol{H}\boldsymbol{r} \tag{3.2}$$

which express the nodal loads $\boldsymbol{p}$ in terms of the element stress-resultants $\boldsymbol{r}$. These equations may be written in the alternative form $\boldsymbol{p}_i = \boldsymbol{H}_i\boldsymbol{r}$, where the suffix i denotes the individual nodes of the element. Any vector $\boldsymbol{p}$ which satisfies (3.2) will maintain the element in overall equilibrium. Finally (provided that $\boldsymbol{r}$ and $\boldsymbol{e}$ correspond in a work sense) we have the compatibility equations

$$\boldsymbol{e} = \boldsymbol{H}^t\boldsymbol{d} \tag{3.3}$$

which relate the deformations $\boldsymbol{e}$ to the nodal displacements $\boldsymbol{d}$. These equations may be written in the form $\boldsymbol{e} = \boldsymbol{H}_j{}^t\boldsymbol{d}_j$, where the suffix j denotes the individual nodes and repetition of the suffix indicates summation over all nodes of the element. The matrix $\boldsymbol{H}$ depends on the geometric form of the element but not on its elastic properties. Equations (3.1), (3.2) and (3.3) may be combined to give the nodal load/displacement equations

$$\boldsymbol{p} = \boldsymbol{Hr} = \boldsymbol{HKe} = \boldsymbol{HKH}^t\boldsymbol{d} \tag{3.4a}$$

which may be written as

$$\boldsymbol{p}_i = \boldsymbol{H}_i\boldsymbol{KH}_j{}^t\boldsymbol{d}_j = \boldsymbol{K}_{ij}\boldsymbol{d}_j. \tag{3.4b}$$

In these two chapters we derive the $\boldsymbol{K}$, $\boldsymbol{H}$ and $\boldsymbol{K}_{ij}$ matrices for a number of different structural elements. In the present chapter we discuss line elements, i.e. beams, columns and arches, while in Chapter 4 we consider some of the simpler elements used in continuum analysis. In each chapter we begin by assuming that any applied loads acting at internal points of elements have been replaced by equivalent nodal loads. Techniques for finding these loads are given at the ends of the chapters. Although the layout of the two chapters is similar there are basic differences between the two classes of element which have a considerable effect on the analysis.

The most important line element for the structural engineer is the uniform straight beam. In the first part of the chapter we develop the $\boldsymbol{K}_{ij}$ matrices for this element and consider a number of special cases which are of practical importance. The numerical results we derive will be familiar to most readers, since the nodal load/displacement equations are simply the slope/deflection equations of classical structural analysis. However, the approach via the $\boldsymbol{K}$ and $\boldsymbol{H}$ matrices provides an interesting alternative to the traditional procedure.

Following the discussion of straight uniform members we derive a general expression for the flexibility matrix $\boldsymbol{F}$ of a curved beam of non-uniform section. An extension of this analysis gives an expression for the flexibility matrix of a segmented member in terms of the flexibility matrices of its component parts. This provides a convenient method of analysing members with various types of joint connection. Finally we return to the problem postponed from the start of the chapter—the calculation of equivalent joint loads. In static problems these may be due either to loads acting at internal points of members or to initial strains (i.e. temperature effects). In dynamic problems they arise from the distributed mass of the members.

For the sake of simplicity we devote most of our attention to plane members. However, the general results we derive are also applicable to three-dimensional members, provided that the various vectors and matrices are defined appropriately. In view of its practical importance the extension of the theory for a straight uniform beam to three dimensions is given in full.

Throughout our treatment of line elements we shall make the usual assumptions of elementary bending theory, viz.

1. All displacements are small compared with the dimensions of the element.
2. Plane sections remain plane.
3. Direct stresses normal to the centre-line of the element are ignored.
4. Shear strains are ignored.

With these assumptions the deformation of a line element is completely defined by the displacement of its centre-line. Since a line element has two nodes, equations (3.4b) may be written out in full as

$$\left.\begin{aligned} \boldsymbol{p}_1 &= \boldsymbol{K}_{11}\boldsymbol{d}_1 + \boldsymbol{K}_{12}\boldsymbol{d}_2, \\ \boldsymbol{p}_2 &= \boldsymbol{K}_{21}\boldsymbol{d}_1 + \boldsymbol{K}_{22}\boldsymbol{d}_2. \end{aligned}\right\} \qquad (3.5)$$

3.1. Stress-resultant and deformation vectors for a general plane member

Although our primary objective at this stage is to derive equations for a straight uniform member, the vectors $\boldsymbol{r}$ and $\boldsymbol{e}$ are defined in a similar way for both straight and curved members. The equilibrium matrices $\boldsymbol{H}_1$, $\boldsymbol{H}_2$ are also independent of the shape and elastic properties of a member. We therefore begin our analysis by considering a general plane member loaded only by forces and moments at its ends.

Such a member is shown in Fig. 3.1a, the origin O and the directions of the axes x, y being chosen arbitrarily. (We shall see later that in particular cases a suitable choice of axes greatly simplifies the analysis.) The nodal loads and nodal displacements are the vectors

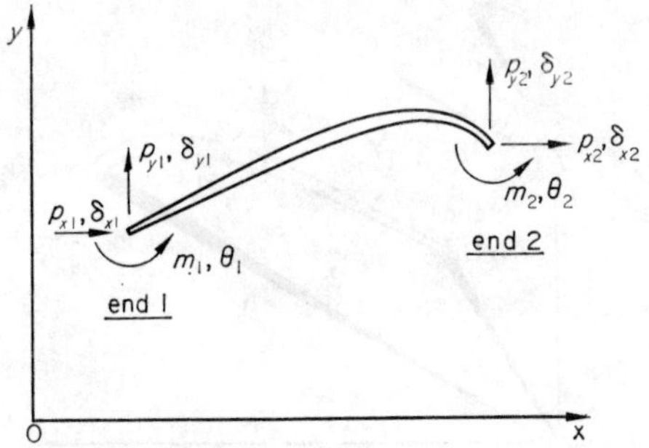

FIG. 3.1a. End-loads and end-displacements for a general plane member.

$$\boldsymbol{p} = \begin{bmatrix} \boldsymbol{p}_1 \\ \boldsymbol{p}_2 \end{bmatrix} = \begin{bmatrix} p_{x1} \\ p_{y1} \\ m_1 \\ p_{x2} \\ p_{y2} \\ m_2 \end{bmatrix} \qquad \boldsymbol{d} = \begin{bmatrix} \boldsymbol{d}_1 \\ \boldsymbol{d}_2 \end{bmatrix} = \begin{bmatrix} \delta_{x1} \\ \delta_{y1} \\ \theta_1 \\ \delta_{x2} \\ \delta_{y2} \\ \theta_2 \end{bmatrix}$$

Since there are six components in the vectors $\boldsymbol{p}$ and $\boldsymbol{d}$ and three equations of equilibrium there are three components in each of the vectors $\boldsymbol{r}$ and $\boldsymbol{e}$. As mentioned in the previous chapter, either $\boldsymbol{r}$ or $\boldsymbol{e}$ may be chosen arbitrarily. We choose $\boldsymbol{r}$ as follows. We imagine that two rigid arms $O1$, $O2$ are attached to the ends of the member, as shown in Fig. 3.1b, and that a self-equilibrating load-pair $-\boldsymbol{r}$, $\boldsymbol{r}$ is applied to the ends of the two arms at O, where

$$\boldsymbol{r} = \begin{bmatrix} t \\ q \\ m \end{bmatrix}$$

as shown in the figure. The stresses at any point in the member may now be expressed in terms of $\boldsymbol{r}$. If p_{xS}, p_{yS} and m_S are the components of force and the bending moment acting on the *left-hand side* of a section through the beam at the point S in Fig. 3.1b it follows that

$$\begin{bmatrix} p_{xS} \\ p_{yS} \\ m_S \end{bmatrix} = \begin{bmatrix} 1 & 0 & 0 \\ 0 & 1 & 0 \\ y_S & -x_S & 1 \end{bmatrix} \begin{bmatrix} t \\ q \\ m \end{bmatrix}$$

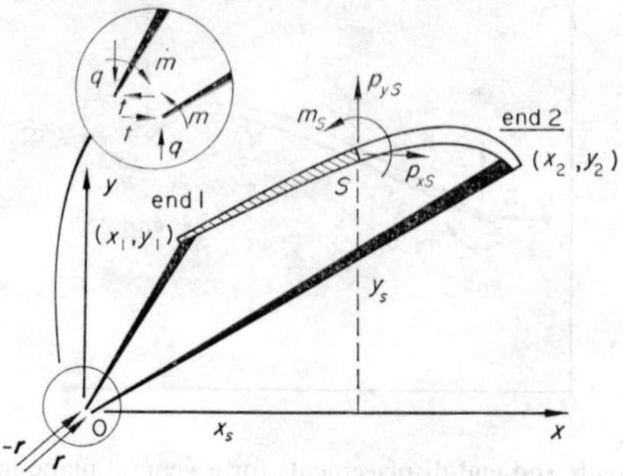

FIG. 3.1b. The stress-resultant $\boldsymbol{r}$ for a general plane member.

which may be written as $\boldsymbol{p}_S = \boldsymbol{H}_S\boldsymbol{r}$. The loads $\boldsymbol{p}_1$ and $\boldsymbol{p}_2$ applied to the ends of the member proper are given by

$$\begin{bmatrix} p_{x1} \\ p_{y1} \\ m_1 \end{bmatrix} = -\begin{bmatrix} 1 & 0 & 0 \\ 0 & 1 & 0 \\ y_1 & -x_1 & 1 \end{bmatrix}\begin{bmatrix} t \\ q \\ m \end{bmatrix}, \quad \begin{bmatrix} p_{x2} \\ p_{y2} \\ m_2 \end{bmatrix} = \begin{bmatrix} 1 & 0 & 0 \\ 0 & 1 & 0 \\ y_2 & -x_2 & 1 \end{bmatrix}\begin{bmatrix} t \\ q \\ m \end{bmatrix} \tag{3.6a}$$

or

$$\boldsymbol{p}_1 = \boldsymbol{H}_1\boldsymbol{r}, \quad \boldsymbol{p}_2 = \boldsymbol{H}_2\boldsymbol{r}. \tag{3.6b}$$

Note the negative sign which appears in the equation for $\boldsymbol{p}_1$ and in the definition of $\boldsymbol{H}_1$. If S coincides with end 1 then $\boldsymbol{p}_S = -\boldsymbol{p}_1$.

The vector $\boldsymbol{e}$ which corresponds to $\boldsymbol{r}$ is the *relative* displacement of the ends of the two arms at O, that is,

$$\boldsymbol{e} = \begin{bmatrix} e_x \\ e_y \\ \phi \end{bmatrix}$$

where e_x, e_y are the two components of relative translation and ϕ is the relative rotation. (This follows from the fact that in a virtual deformation $\boldsymbol{e}^*$ the work done on the member must be $\boldsymbol{r}^t\boldsymbol{e}^*$.) It is easy to verify, either by a virtual work argument or by direct geometrical reasoning, that

$$\boldsymbol{e} = \boldsymbol{H}_1{}^t\boldsymbol{d}_1 + \boldsymbol{H}_2{}^t\boldsymbol{d}_2. \tag{3.7}$$

We may note in passing that equation (3.7) gives us the condition $\boldsymbol{H}_1'\boldsymbol{d}_1 + \boldsymbol{H}_2'\boldsymbol{d}_2 = \boldsymbol{0}$ for a purely rigid-body displacement.

Equations (3.6) and (3.7) hold whatever the shape of a member, and the extension of the whole analysis to three dimensions is straightforward. The next step is to determine the form of $\boldsymbol{K}$ in some specific cases.

3.2. The straight uniform beam: two-dimensional analysis

When dealing with a straight uniform beam it is convenient to locate the origin at the mid-point of the beam and make the x-axis coincide with the undeformed centre-line. We adopt the convention (which will be used throughout the book) that the direction end 1 → end 2 defines the positive direction of x. With this coordinate system we have

$$\boldsymbol{H}_1 = -\begin{bmatrix} 1 & 0 & 0 \\ 0 & 1 & 0 \\ 0 & L/2 & 1 \end{bmatrix}, \qquad \boldsymbol{H}_2 = \begin{bmatrix} 1 & 0 & 0 \\ 0 & 1 & 0 \\ 0 & -L/2 & 0 \end{bmatrix}$$

This choice of coordinate system leads to a very simple relationship between $\boldsymbol{r}$ and $\boldsymbol{e}$. The component t produces a purely axial force in the beam and is related to the extension e_x by the equation

$$t = \frac{EA}{L}\, e_x. \tag{3.8a}$$

The component q produces a constant shear force and the deformed shape of the beam is shown in Fig. 3.2a. By symmetry there is no tendency for relative rotation of the two arms. The relationship between q and e_y is

$$q = \frac{12EI}{L^3}\, e_y. \tag{3.8b}$$

This result may be found in any textbook on elementary structural theory, or may be derived by integrating the differential equation of bending with appropriate boundary conditions. Finally, the component m produces a constant moment, the deformed shape being shown in Fig. 3.2b. Again by symmetry there is no tendency to relative translation of the two arms at

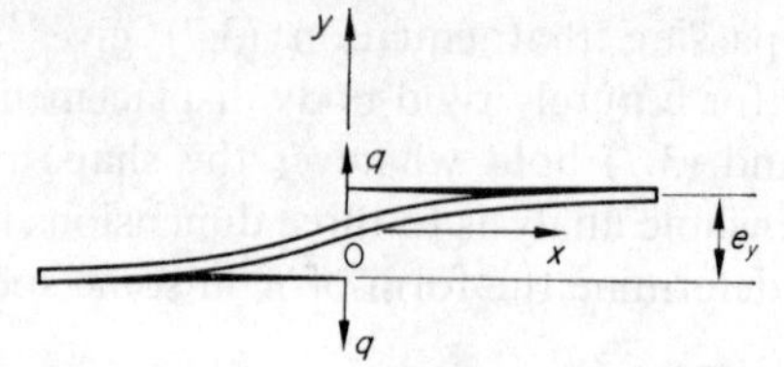

FIG. 3.2a. Deformation of a straight uniform member associated with q.

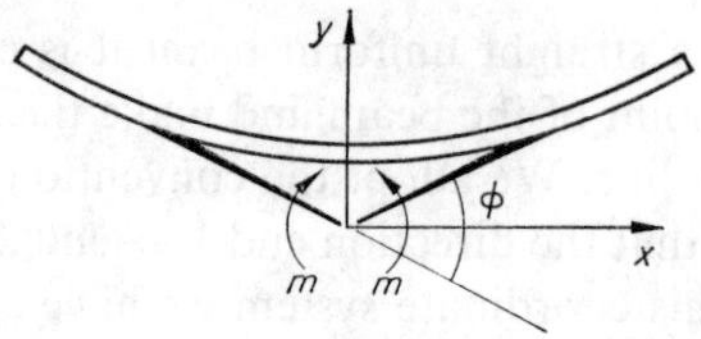

FIG. 3.2b. Deformation of a straight uniform member associated with m.

O. The relationship between m and ϕ is

$$m = \frac{EI}{L}\,\phi. \tag{3.8c}$$

Collecting these three equations we obtain

$$\begin{bmatrix} t \\ q \\ m \end{bmatrix} = \begin{bmatrix} EA/L & 0 & 0 \\ 0 & 12EI/L^3 & 0 \\ 0 & 0 & EI/L \end{bmatrix} \begin{bmatrix} e_x \\ e_y \\ \phi \end{bmatrix}, \text{ or } \boldsymbol{r} = \boldsymbol{Ke} \tag{3.9a}$$

Inverting the matrix $\boldsymbol{K}$ we obtain

$$\begin{bmatrix} e_x \\ e_y \\ \phi \end{bmatrix} = \begin{bmatrix} L/EA & 0 & 0 \\ 0 & L^3/12EI & 0 \\ 0 & 0 & L/EI \end{bmatrix} \begin{bmatrix} t \\ q \\ m \end{bmatrix}, \text{ or } \boldsymbol{e} = \boldsymbol{Fr} \tag{3.9b}$$

We now combine (3.6b) and (3.7) with (3.9a) to obtain

$$\left.\begin{aligned} \boldsymbol{p}_1 &= \boldsymbol{H}_1\boldsymbol{r} = \boldsymbol{H}_1\boldsymbol{Ke} = \boldsymbol{H}_1\boldsymbol{KH}_1{}^t\boldsymbol{d}_1 + \boldsymbol{H}_1\boldsymbol{KH}_2{}^t\boldsymbol{d}_2 \\ \boldsymbol{p}_2 &= \boldsymbol{H}_2\boldsymbol{r} = \boldsymbol{H}_2\boldsymbol{Ke} = \boldsymbol{H}_2\boldsymbol{KH}_1{}^t\boldsymbol{d}_1 + \boldsymbol{H}_2\boldsymbol{KH}_2{}^t\boldsymbol{d}_2 \end{aligned}\right\} \tag{3.10a}$$

or

$$\left.\begin{aligned} \boldsymbol{p}_1 &= \boldsymbol{K}_{11}\boldsymbol{d}_1 + \boldsymbol{K}_{12}\boldsymbol{d}_2 \\ \boldsymbol{p}_2 &= \boldsymbol{K}_{21}\boldsymbol{d}_1 + \boldsymbol{K}_{22}\boldsymbol{d}_2 \end{aligned}\right\} \tag{3.10b}$$

where $K_{ij} = H_i K H_j^t$. Evaluating the K_{ij} matrices from the H and K matrices defined above gives

$$\left.\begin{aligned} K_{11} &= \begin{bmatrix} EA/L & 0 & 0 \\ 0 & 12EI/L^3 & 6EI/L^2 \\ 0 & 6EI/L^2 & 4EI/L \end{bmatrix}, \quad K_{12} = \begin{bmatrix} -EA/L & 0 & 0 \\ 0 & -12EI/L^3 & 6EI/L^2 \\ 0 & -6EI/L^2 & 2EI/L \end{bmatrix} \\ K_{21} &= \begin{bmatrix} -EA/L & 0 & 0 \\ 0 & -12EI/L^3 & -6EI/L^2 \\ 0 & 6EI/L^2 & 2EI/L \end{bmatrix}, \quad K_{22} = \begin{bmatrix} EA/L & 0 & 0 \\ 0 & 12EI/L^3 & -6EI/L^2 \\ 0 & -6EI/L^2 & 4EI/L \end{bmatrix} \end{aligned}\right\} \tag{3.11}$$

The load/displacement equations (3.10b) may also be derived from the differential equations of bending. Alternatively they may be derived by a similar analysis to that given above, but using a different stress-resultant vector r. In that case the K and H matrices would be different, but the final matrices K_{11}, etc., would be the same. Note that the equations are symmetric (K_{11} and K_{12} are symmetric, and $K_{21} = K_{12}^t$). They are also singular, since the beam may be given a (small) arbitrary rigid-body displacement without affecting the end-loads p_1 and p_2.

The procedure for transforming the load/displacement equations into global coordinates follows the pattern set out in Section 2.4. The force components p_x, p_y and the translation components δ_x, δ_y transform in the manner described in that section, while the moments and rotations are unaffected by the change of axes. Thus we have

$$p_i' = Tp_i, \; d_i = T^t d_i', \quad i = 1,2$$

where

$$T = \begin{bmatrix} \cos\alpha & -\sin\alpha & 0 \\ \sin\alpha & \cos\alpha & 0 \\ 0 & 0 & 1 \end{bmatrix} \tag{3.12}$$

This relationship may be verified from Fig. 3.3. Substituting for the vectors p_i and d_i the load/displacement equations (3.10b) become

$$\begin{aligned} p_1' &= TK_{11}T^t d_1' + TK_{12}T^t d_2', \\ p_2' &= TK_{21}T^t d_1' + TK_{22}T^t d_2' \end{aligned}$$

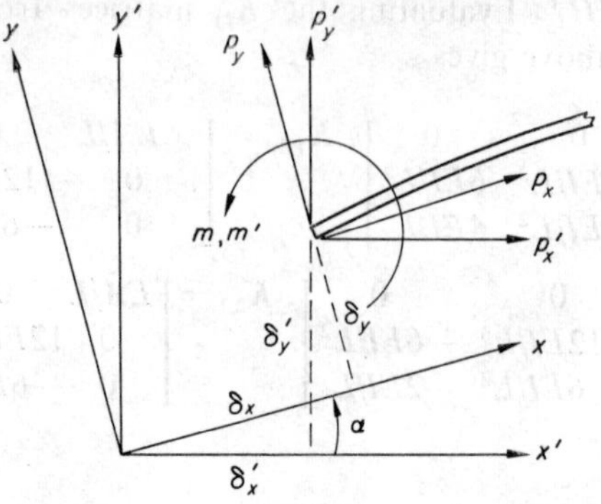

FIG. 3.3. Change of coordinates for a plane member.

or

$$\left.\begin{aligned} \boldsymbol{p}_1' &= \boldsymbol{K}_{11}'\boldsymbol{d}_1' + \boldsymbol{K}_{12}'\boldsymbol{d}_2', \\ \boldsymbol{p}_2' &= \boldsymbol{K}_{21}'\boldsymbol{d}_1' + \boldsymbol{K}_{22}'\boldsymbol{d}_2' \end{aligned}\right\} \tag{3.13a}$$

where $\boldsymbol{K}_{ij}' = \boldsymbol{T}\boldsymbol{K}_{ij}\boldsymbol{T}^t$. Equations (3.13a), like (3.10), are symmetric and singular. The matrix $\boldsymbol{K}_{11}'$ has the form

$$\begin{bmatrix} C^2EA/L + S^2 12EI/L^3 & SC(EA/L - 12EI/L^3) & -S\,6EI/L^2 \\ SC(EA/L - 12EI/L^3) & S^2EA/L + C^2 12EI/L^3 & C\,6EI/L^2 \\ -S\,6EI/L^2 & C\,6EI/L^2 & 4EI/L \end{bmatrix} \tag{3.13b}$$

where $S = \sin\alpha$, $C = \cos\alpha$. The other matrices are similar.

3.3. The straight uniform beam: three-dimensional analysis

The development of the equations for a beam in three dimensions follows very easily from the work of the previous section. Such a beam is shown in Fig. 3.4. It is assumed that the *y*- and *z*-axes are the principal axes of inertia, and that the flexural rigidities for bending about these axes are EI_y and EI_z respectively. Once again we take the origin of co-ordinates at the mid-point of the beam.

There are now six components of load and displacement at each end—12 in all. There are six equations of equilibrium, so that the vectors $\boldsymbol{r}$ and $\boldsymbol{e}$ each have six components. If we look at the previous analysis the

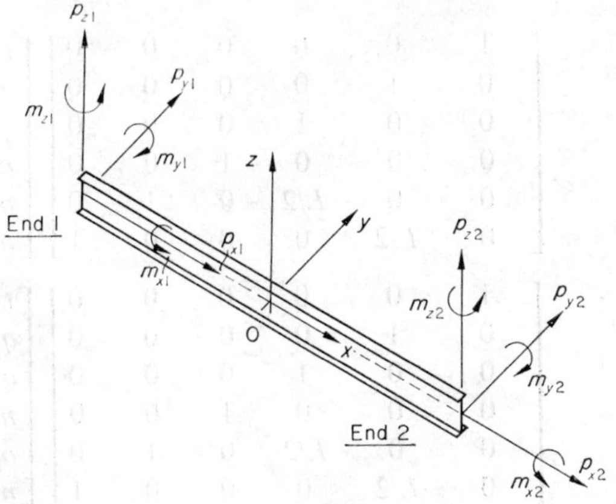

FIG. 3.4. Coordinate axes and end-loads for a three-dimensional member.

additional components of $\boldsymbol{r}$ are easily found. Each of the principal planes of bending xy and xz requires a shear force component and a moment component. The components associated with the xy plane are q_y and m_z, while the components associated with the xz plane are q_z and m_y. We also require a torque m_x which, like the tension and the shear forces, is constant along the length of the beam. If we consider only the simple case in which warping effects are ignored and assume the torsion axis of the beam to coincide with the centre-line (i.e. the x-axis) then the equation for twisting is simply

$$m_x = (GJ/L)\phi_x$$

where ϕ_x is the amount of twist in the beam. Thus equation (3.9a) becomes

$$\begin{bmatrix} t \\ q_y \\ q_z \\ m_x \\ m_y \\ m_z \end{bmatrix} = \begin{bmatrix} EA/L & 0 & 0 & 0 & 0 & 0 \\ 0 & 12EI_z/L^3 & 0 & 0 & 0 & 0 \\ 0 & 0 & 12EI_y/L^3 & 0 & 0 & 0 \\ 0 & 0 & 0 & GJ/L & 0 & 0 \\ 0 & 0 & 0 & 0 & EI_y/L & 0 \\ 0 & 0 & 0 & 0 & 0 & EI_z/L \end{bmatrix} \begin{bmatrix} e_x \\ e_y \\ e_z \\ \phi_x \\ \phi_y \\ \phi_z \end{bmatrix} \quad (3.14)$$

This equation defines $\boldsymbol{K}$. It is easy to verify that the equilibrium equations are

$$
\left.
\begin{aligned}
\begin{bmatrix} p_{x1} \\ p_{y1} \\ p_{z1} \\ m_{x1} \\ m_{y1} \\ m_{z1} \end{bmatrix} &= -\begin{bmatrix} 1 & 0 & 0 & 0 & 0 & 0 \\ 0 & 1 & 0 & 0 & 0 & 0 \\ 0 & 0 & 1 & 0 & 0 & 0 \\ 0 & 0 & 0 & 1 & 0 & 0 \\ 0 & 0 & -L/2 & 0 & 1 & 0 \\ 0 & L/2 & 0 & 0 & 0 & 1 \end{bmatrix} \begin{bmatrix} t \\ q_y \\ q_z \\ m_x \\ m_y \\ m_z \end{bmatrix} \\
\begin{bmatrix} p_{x2} \\ p_{y2} \\ p_{z2} \\ m_{x2} \\ m_{y2} \\ m_{z2} \end{bmatrix} &= \begin{bmatrix} 1 & 0 & 0 & 0 & 0 & 0 \\ 0 & 1 & 0 & 0 & 0 & 0 \\ 0 & 0 & 1 & 0 & 0 & 0 \\ 0 & 0 & 0 & 1 & 0 & 0 \\ 0 & 0 & L/2 & 0 & 1 & 0 \\ 0 & -L/2 & 0 & 0 & 0 & 1 \end{bmatrix} \begin{bmatrix} t \\ q_y \\ q_z \\ m_x \\ m_y \\ m_z \end{bmatrix}
\end{aligned}
\right\} \quad (3.15)
$$

These equations define $\boldsymbol{H}_1$ and $\boldsymbol{H}_2$. Substituting the $\boldsymbol{K}$ and $\boldsymbol{H}$ matrices given in (3.14) and (3.15) into the expression $\boldsymbol{K}_{ij} = \boldsymbol{H}_i\boldsymbol{K}\boldsymbol{H}_j{}^t$ gives

$$
\left.
\begin{aligned}
\boldsymbol{K}_{11} &= \begin{bmatrix} EA/L & 0 & 0 & 0 & 0 & 0 \\ 0 & 12EI_z/L^3 & 0 & 0 & 0 & 6EI_z/L^2 \\ 0 & 0 & 12EI_y/L^3 & 0 & -6EI_y/L^2 & 0 \\ 0 & 0 & 0 & GJ/L & 0 & 0 \\ 0 & 0 & -6EI_y/L^2 & 0 & 4EI_y/L & 0 \\ 0 & 6EI_z/L^2 & 0 & 0 & 0 & 4EI_z/L \end{bmatrix} \\
\boldsymbol{K}_{12} = \boldsymbol{K}_{21}{}^t &= \begin{bmatrix} -EA/L & 0 & 0 & 0 & 0 & 0 \\ 0 & -12EI_z/L^3 & 0 & 0 & 0 & 6EI_z/L^2 \\ 0 & 0 & -12EI_y/L^3 & 0 & -6EI_y/L^2 & 0 \\ 0 & 0 & 0 & -GJ/L & 0 & 0 \\ 0 & 0 & 6EI_y/L^2 & 0 & 2EI_y/L & 0 \\ 0 & -6EI_z/L^2 & 0 & 0 & 0 & 2EI_z/L \end{bmatrix}
\end{aligned}
\right\} \quad (3.16)
$$

$\boldsymbol{K}_{22}$ being equal to $\boldsymbol{K}_{11}$ with the signs of the off-diagonal elements $-6EI_y/L^2$ and $6EI_z/L^2$ reversed.

In three dimensions moments and (small) rotations transform in the same way as forces and translations. The transformation equations for loads are

$$\begin{bmatrix} p_x' \\ p_y' \\ p_z' \\ m_x' \\ m_y' \\ m_z' \end{bmatrix} = \begin{bmatrix} \widehat{x'x} & \widehat{x'y} & \widehat{x'z} & 0 & 0 & 0 \\ \widehat{y'x} & \widehat{y'y} & \widehat{y'z} & 0 & 0 & 0 \\ \widehat{z'x} & \widehat{z'y} & \widehat{z'z} & 0 & 0 & 0 \\ 0 & 0 & 0 & \widehat{x'x} & \widehat{x'y} & \widehat{x'z} \\ 0 & 0 & 0 & \widehat{y'x} & \widehat{y'y} & \widehat{y'z} \\ 0 & 0 & 0 & \widehat{z'x} & \widehat{z'y} & \widehat{z'z} \end{bmatrix} \begin{bmatrix} p_x \\ p_y \\ p_z \\ m_x \\ m_y \\ m_z \end{bmatrix} \quad (3.17)$$

where $\widehat{x'x}$, $\widehat{x'y}$, etc., are the cosines of the angles between the specified axes. We write this in the usual way as $\boldsymbol{p}' = \boldsymbol{Tp}$. The transformation of the load/displacement equations into global coordinates now follows in exactly the same way as for a member in a plane. Equations (3.13a) still hold, provided that the appropriate three-dimensional forms of the $\boldsymbol{K}$ and $\boldsymbol{T}$ matrices are used. The computation of the matrices $\boldsymbol{K}_{ij}' = \boldsymbol{TK}_{ij}\boldsymbol{T}^t$ is simplified by partitioning the matrices $\boldsymbol{K}_{ij}$ and $\boldsymbol{T}$ into 3×3 sub-matrices, writing them in the form

$$\boldsymbol{K}_{ij} = \begin{bmatrix} \boldsymbol{K}_\alpha & \boldsymbol{K}_\beta \\ \boldsymbol{K}_\gamma & \boldsymbol{K}_\delta \end{bmatrix}_{ij} \qquad \boldsymbol{T} = \begin{bmatrix} \boldsymbol{T}_0 & \boldsymbol{0} \\ \boldsymbol{0} & \boldsymbol{T}_0 \end{bmatrix}$$

It is easy to show that

$$\boldsymbol{K}_{ij}' = \boldsymbol{TK}_{ij}\boldsymbol{T}^t$$

$$= \begin{bmatrix} \boldsymbol{T}_0 & \boldsymbol{0} \\ \boldsymbol{0} & \boldsymbol{T}_0 \end{bmatrix} \begin{bmatrix} \boldsymbol{K}_\alpha & \boldsymbol{K}_\beta \\ \boldsymbol{K}_\gamma & \boldsymbol{K}_\delta \end{bmatrix}_{ij} \begin{bmatrix} \boldsymbol{T}_0^t & \boldsymbol{0} \\ \boldsymbol{0} & \boldsymbol{T}_0^t \end{bmatrix} = \begin{bmatrix} \boldsymbol{T}_0\boldsymbol{K}_\alpha\boldsymbol{T}_0^t & \boldsymbol{T}_0\boldsymbol{K}_\beta\boldsymbol{T}_0^t \\ \boldsymbol{T}_0\boldsymbol{K}_\gamma\boldsymbol{T}_0^t & \boldsymbol{T}_0\boldsymbol{K}_\delta\boldsymbol{T}_0^t \end{bmatrix}_{ij}$$

Thus triple matrix products involving 6×6 matrices are replaced by similar products involving 3×3 matrices. Since the number of operations in a matrix multiplication is proportional to the cube of the size of the matrices this represents a considerable saving in computational effort.

3.4. Straight uniform members with pinned ends

The stiffness matrices for pin-ended members in a plane or in space may be deduced by setting $I = 0$ in the matrices already derived for a beam and dropping all coefficients associated with moments and rotations.

Alternatively we may start from equations (2.2)

$$\left.\begin{aligned} p_1 &= Kd_1 - Kd_2 \\ p_2 &= -Kd_1 + Kd_2 \end{aligned}\right\} \tag{3.18}$$

where $K = EA/L$. These equations are in member coordinates. From Fig. 3.5 we have

$$\boldsymbol{p}' = \begin{bmatrix} \cos \alpha \\ \sin \alpha \end{bmatrix} p, \quad d = [\cos \alpha \quad \sin \alpha]\, \boldsymbol{d}' \qquad (3.19a), (3.19b)$$

which we may write in the usual way as $\boldsymbol{p}' = \boldsymbol{T}p$, $d = \boldsymbol{T}^t\boldsymbol{d}'$. Substituting for the p's and d's in (3.18) we obtain

$$\left.\begin{aligned} \boldsymbol{p}' &= \boldsymbol{K}'\boldsymbol{d}_1' - \boldsymbol{K}'\boldsymbol{d}_2' \\ \boldsymbol{p}' &= -\boldsymbol{K}'\boldsymbol{d}_1' + \boldsymbol{K}'\boldsymbol{d}_2' \end{aligned}\right\} \tag{3.20a}$$

where

$$\boldsymbol{K}' = \boldsymbol{TKT}^t = \frac{EA}{L}\begin{bmatrix} \cos^2\alpha & \cos \alpha \sin \alpha \\ \cos \alpha \sin \alpha & \sin^2\alpha \end{bmatrix} \tag{3.20b}$$

This agrees with the result of setting $I = 0$ in (3.13b) and dropping the last row and column. Note that since the matrix $\boldsymbol{T}$ in (3.19a) is not a square matrix it is not possible to "invert" $\boldsymbol{T}$. However, equation (3.19a) may be pre-multiplied by $[\cos \alpha \ \sin \alpha]$ to give $p = [\cos \alpha \ \sin \alpha]\, \boldsymbol{p}'$, which may be written as $p = \boldsymbol{T}^t\boldsymbol{p}'$. The same procedure cannot be applied to (3.19b).

In three dimensions the vectors $\boldsymbol{p}'$ and $\boldsymbol{d}'$ each have three components

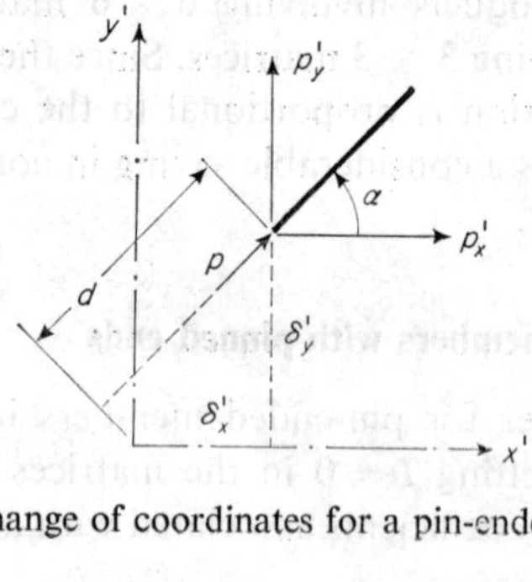

FIG. 3.5. Change of coordinates for a pin-ended member.

and equations (3.19) become

$$\boldsymbol{p}' = \begin{bmatrix} l \\ m \\ n \end{bmatrix} p, \qquad d = [l \quad m \quad n]\,\boldsymbol{d}' \qquad \text{(3.19c), (3.19d)}$$

where l, m, n are the direction cosines of the member with respect to the global axes.

Equation (3.20a) still holds, with

$$\boldsymbol{K}' = \boldsymbol{TKT}^t = \frac{EA}{L}\begin{bmatrix} l^2 & lm & ln \\ lm & m^2 & mn \\ ln & nm & n^2 \end{bmatrix} \tag{3.20c}$$

Note that for pin-ended members the convention that the direction end 1→end 2 coincides with the positive x-axis is not necessary. The equations are the same if ends 1 and 2 are reversed. Also the matrices $\boldsymbol{H}_1$, $\boldsymbol{H}_2$ degenerate into scalar multipliers with values -1 and 1 respectively.

3.5. Straight uniform members in plane grillages

A member bending in a plane and a pin-ended member can both be regarded as special cases of the three-dimensional member considered in Section 3.3. Another important special case is a member of a plane grillage. Grillages are often complex highly redundant structures and matrix methods are therefore useful in their analysis.

We consider a straight uniform member of a plane horizontal grillage, as shown in Fig. 3.6. We imagine that the grillage is loaded by vertical loads at the joints and (possibly) by external moments about the x- and y-axes. Because of these restrictions on the loading we neglect translations of the joints in the xy plane and rotations about the z-axis. As in Section 3.3, we neglect warping effects and assume that the beam twists about the x-axis.

These assumptions reduce the member to a beam bending in the xz plane, with torsion about the x-axis rather than axial force along it. The load and displacement vectors thus have components

$$\boldsymbol{p} = \begin{bmatrix} m_x \\ p_z \\ m_y \end{bmatrix} \qquad \boldsymbol{d} = \begin{bmatrix} \theta_x \\ \delta_z \\ \theta_y \end{bmatrix}$$

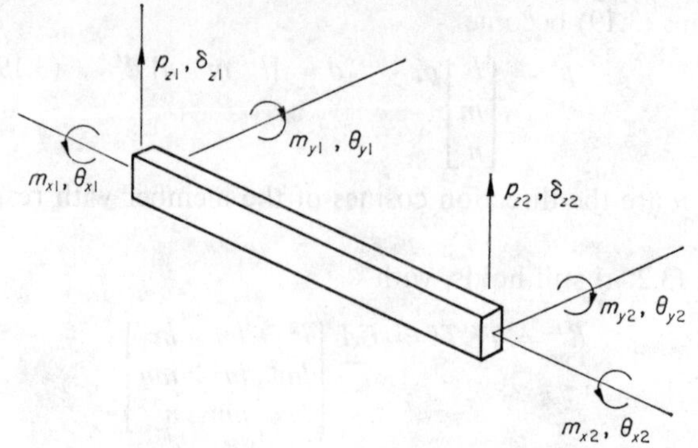

FIG. 3.6. A straight uniform member in a plane grillage.

and the $\boldsymbol{K}_{ij}$ matrices are

$$\boldsymbol{K}_{11} = \begin{bmatrix} GJ/L & 0 & 0 \\ 0 & 12EI/L^3 & 6EI/L^2 \\ 0 & 6EI/L^2 & 4EI/L \end{bmatrix} \quad \boldsymbol{K}_{12} = \begin{bmatrix} -GJ/L & 0 & 0 \\ 0 & -12EI/L^3 & 6EI/L^2 \\ 0 & -6EI/L^2 & 2EI/L \end{bmatrix}$$

$$\boldsymbol{K}_{21} = \begin{bmatrix} -GJ/L & 0 & 0 \\ 0 & -12EI/L^3 & -6EI/L^2 \\ 0 & 6EI/L^2 & 2EI/L \end{bmatrix} \quad \boldsymbol{K}_{22} = \begin{bmatrix} -GJ/L & 0 & 0 \\ 0 & 12EI/L^3 & -6EI/L^2 \\ 0 & -6EI/L^2 & 4EI/L \end{bmatrix}$$

Although these matrices are very similar to those obtained for a plane beam in bending, the coordinate transformation matrix $\boldsymbol{T}$ is different. From Fig. 3.7 we have

$$\begin{bmatrix} m_x' \\ p_z' \\ m_y' \end{bmatrix} = \begin{bmatrix} \cos\alpha & 0 & -\sin\alpha \\ 0 & 1 & 0 \\ \sin\alpha & 0 & \cos\alpha \end{bmatrix} \begin{bmatrix} m_x \\ p_z \\ m_y \end{bmatrix}$$

which we write in the normal way as $\boldsymbol{p}' = \boldsymbol{T}\boldsymbol{p}$. The reader may easily verify that

$$\boldsymbol{K}_{11}' = \begin{bmatrix} C^2\,GJ/L + S^2\,4EI/L & -S\,6EI/L^2 & SC(GJ/L - 4EI/L) \\ -S\,6EI/L^2 & 12EI/L^3 & C\,6EI/L^2 \\ SC(GJ/L - 4EI/L) & C\,6EI/L^2 & S^2GJ/L + C^2\,4EI/L \end{bmatrix}$$

where $S = \sin\alpha$, $C = \cos\alpha$. The other $\boldsymbol{K}_{ij}'$ matrices are similar.

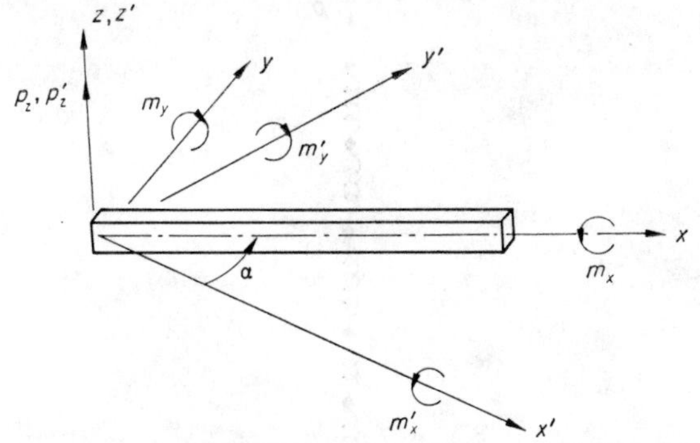

FIG. 3.7. Change of coordinates for a member in a plane grillage.

3.6. Curved and non-uniform plane members

We now consider the problem of finding the matrix $\boldsymbol{K}$ of a member of arbitrary shape and variable cross-section. We develop the analysis for a plane member, although the final result holds quite generally provided that the relevant matrices are defined appropriately.

Before commencing the analysis we consider a simpler problem—the one-dimensional chain of springs shown in Fig. 3.8. In this example it is clear that the force in each spring is the same. If we write the load/displacement equations for each spring in the form

extension = flexibility × force

then we can add up the individual extensions to obtain

total extension = sum of flexibilities × force.

This result suggests that the flexibility matrix of a curved non-uniform member might be simply the sum of the flexibility matrices of a chain of elemental segments, each of length ds. We shall see that this is indeed the case, provided that we measure the flexibilities of the segments at a common point and in a common coordinate system.

We consider the member shown in Fig. 3.9. We assume that the centre-line of the member is defined parametrically as $x = x(s)$, $y = y(s)$. Following the example of the chain of springs, we first investigate the contribution which an isolated segment ds makes to the overall deforma-

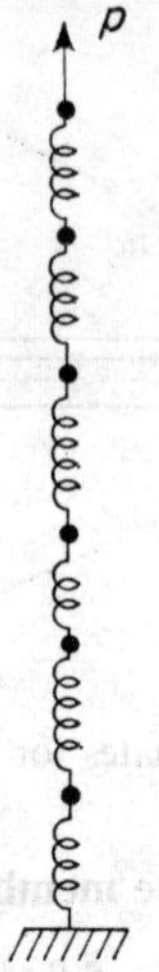

FIG. 3.8. A chain of springs in series.

tion. In the local coordinate system $O_s\hat{x}\hat{y}$ the flexibility matrix of the segment is given by equation (3.9b) as

$$\begin{bmatrix} ds/EA & 0 & 0 \\ 0 & (ds)^3/12EI & 0 \\ 0 & 0 & ds/EI \end{bmatrix}$$

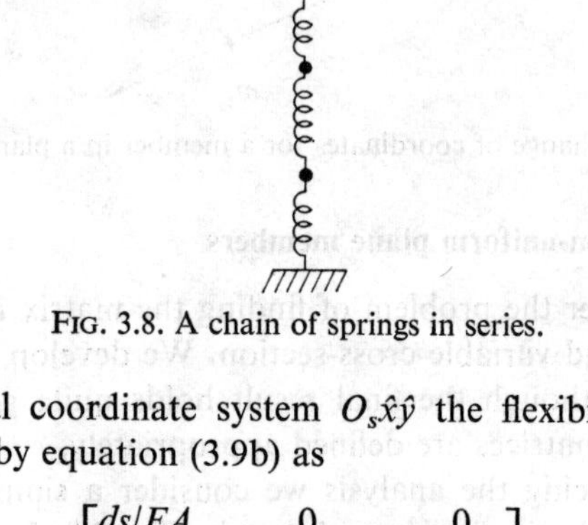

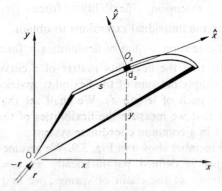

FIG. 3.9. Member and segment coordinate systems for a general plane member.

where A and I are assumed to be known functions of s. We write this as

$$\begin{bmatrix} 1/EA & 0 & 0 \\ 0 & 0 & 0 \\ 0 & 0 & 1/EI \end{bmatrix} ds = \hat{\boldsymbol{F}}_s ds$$

with the "shear" coefficient being of order $(ds)^2$ and therefore disappearing. (Deformation due directly to shear strain may be included at this point if required.)

The self-equilibrating pair of loads $-\boldsymbol{r}$, $\boldsymbol{r}$ applied at the origin $\boldsymbol{O}$ is statically equivalent to loads $-\boldsymbol{r}_s$, $\boldsymbol{r}_s$ applied at the segment origin $\boldsymbol{O}_s$. The two load vectors are related by the equation

$$\boldsymbol{r}_s = \boldsymbol{H}_s \boldsymbol{r} \tag{3.21a}$$

where

$$\boldsymbol{H}_s = \begin{bmatrix} 1 & 0 & 0 \\ 0 & 1 & 0 \\ y(s) & -x(s) & 1 \end{bmatrix} \tag{3.21b}$$

Although the vector $\boldsymbol{r}_s$ acts at O_s it is still expressed in the coordinate system x, y of the complete member. The corresponding vector in the coordinate system $\hat{x}$, $\hat{y}$ is $\hat{\boldsymbol{r}}_s$, where

$$\boldsymbol{r}_s = \boldsymbol{T}_s \hat{\boldsymbol{r}}_s \text{ and } \boldsymbol{T}_s = \begin{bmatrix} dx/ds & -dy/ds & 0 \\ dy/ds & dx/ds & 0 \\ 0 & 0 & 1 \end{bmatrix}$$

Since $\boldsymbol{T}_s$ is orthogonal we have $\boldsymbol{T}_s^{-1} = \boldsymbol{T}_s^t$, so that $\hat{\boldsymbol{r}}_s = \boldsymbol{T}_s^t \boldsymbol{r}_s$. From (3.21a) we obtain

$$\hat{\boldsymbol{r}}_s = \boldsymbol{T}_s^t \boldsymbol{H}_s \boldsymbol{r}. \tag{3.22}$$

Corresponding to (3.22) we have the relationship

$$d\boldsymbol{e} = \boldsymbol{H}_s^t \boldsymbol{T}_s \, d\hat{\boldsymbol{e}}_s \tag{3.23}$$

between the deformation vector $d\hat{\boldsymbol{e}}_s$ of the segment measured at O_s and the corresponding deformation $d\boldsymbol{e}$ (i.e. the relative displacement of the arms at O). This equation may be derived by a virtual work argument in the usual way. Combining (3.22) and (3.23) with the segment flexibility equation $d\hat{e}_s = \hat{F}_s \, ds \, \hat{r}_s$ we obtain

$$de = (H_s^t T_s \hat{F}_s T_s^t H_s \, ds) \, r = F_{sO} \, ds \, r \tag{3.24}$$

where $F_{sO} \, ds$ represents the flexibility matrix of the segment when measured at O in the coordinate system x, y.

Equation (3.24) gives the deformation at O due to the strain in the segment ds. Adding the contributions from all the other segments we obtain

$$e = \left[\int_0^L H_s^t T_s \hat{F}_s T_s^t H_s \, ds \right] r = \int_0^L F_{sO} \, ds \, r = Fr \tag{3.25}$$

where F is the flexibility matrix of the complete member. Expanding the integrand we obtain

$$F = \begin{bmatrix} \displaystyle\int_0^L \frac{(dx/ds)^2}{EA} + \frac{y^2}{EI} \, ds & \displaystyle\int_0^L \frac{(dx/ds)(dy/ds)}{EA} - \frac{xy}{EI} \, ds & \displaystyle\int_0^L \frac{y}{EI} \, ds \\ & \displaystyle\int_0^L \frac{(dy/ds)^2}{EA} + \frac{x^2}{EI} \, ds & \displaystyle -\int_0^L \frac{x}{EI} \, ds \\ \text{Symmetric} & & \displaystyle\int_0^L \frac{1}{EI} \, ds \end{bmatrix} \tag{3.26}$$

This matrix may be inverted to give K, the K_{ij} matrices being obtained from (3.4b) in the usual way. The matrices H_1, H_2 are given by equation (3.6a).

In this analysis the origin and the directions of the coordinate axes were chosen arbitrarily. If the axial strain terms are ignored it is possible to make F a diagonal matrix by choosing O and the axes x, y in such a way that the integrals

$$\int_0^L (x/EI)\,ds, \quad \int_0^L (y/EI)\,ds, \quad \int_0^L (xy/EI)\,ds$$

all vanish. These axes have physical significance in that they are the principal axes of a line distribution of mass, with intensity $1/EI$ per unit length, along the centre-line of the beam. The point O defined in this way is called the *elastic centre*.

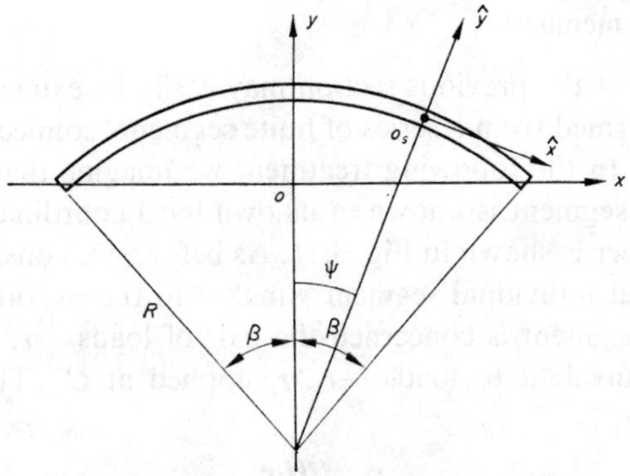

FIG. 3.10. A uniform member forming a circular arc.

As an example of this analysis consider a member of uniform cross-section forming an arc of a circle, as shown in Fig. 3.10. From the figure we have $ds = R d\psi$ and

$$x = R \sin \psi, \quad y = R(\cos \psi - \cos \beta), \quad dx/ds = \cos \psi, \quad dy/ds = -\sin \psi.$$

Evaluating the integrals appearing in (3.26) we obtain

$$\mathbf{F} = \begin{bmatrix} a & 0 & d \\ 0 & b & 0 \\ d & 0 & c \end{bmatrix}$$

where

$$a = \frac{R}{EA}(\beta + \sin\beta\cos\beta) + \frac{R^3}{EI}(\beta(1 + 2\cos^2\beta) - 3\sin\beta\cos\beta),$$

$$b = \frac{R}{EA}(\beta - \sin\beta\cos\beta) + \frac{R^3}{EI}(\beta - \sin\beta\cos\beta),$$

$$c = \frac{2\beta R}{EI}, \qquad d = \frac{2R^2}{EI}(\sin\beta - \beta\cos\beta).$$

3.7. Segmented members

The analysis of the previous section may easily be extended to the case of a member formed from a series of finite segments connected end-to-end by rigid joints. In the following treatment we imagine that the flexibility matrix of each segment is known in its own local coordinate system.

Such a member is shown in Fig. 3.11. As before, we consider the contribution which an individual segment s makes to the overall deformation. As far as this segment is concerned the pair of loads $-\boldsymbol{r}$, $\boldsymbol{r}$ applied at O is statically equivalent to loads $-\boldsymbol{r}_s$, $\boldsymbol{r}_s$ applied at O_s. The relationship between $\boldsymbol{r}$ and $\boldsymbol{r}_s$ is

$$\boldsymbol{r}_s = \boldsymbol{H}_s \boldsymbol{r} \tag{3.27a}$$

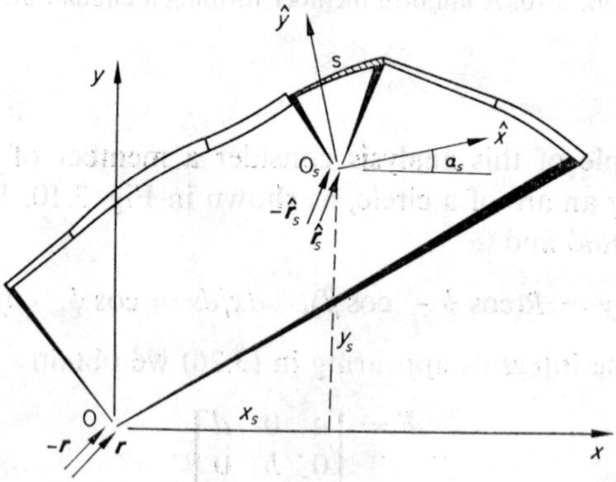

FIG. 3.11. A member comprising a number of segments.

where

$$H_s = \begin{bmatrix} 1 & 0 & 0 \\ 0 & 1 & 0 \\ y_s & -x_s & 1 \end{bmatrix} \qquad (3.27b)$$

The corresponding vector $\hat{r}_s$ in the coordinate system $\hat{x}$, $\hat{y}$ is given by (3.22) as

$$\hat{r}_s = T_s^t H_s r \qquad (3.28)$$

where

$$T_s = \begin{bmatrix} \cos\alpha & -\sin\alpha & 0 \\ \sin\alpha & \cos\alpha & 0 \\ 0 & 0 & 1 \end{bmatrix}$$

Corresponding to (3.28) we have the equation

$$e = H_s^t T_s \hat{e}_s \qquad (3.29)$$

relating the relative displacements at O and O_s. Combining (3.28) and (3.29) with the segment flexibility equation $\hat{e}_s = \hat{F}_s \hat{r}_s$ we obtain

$$e = H_s^t T_s \hat{F}_s T_s^t H_s r = F_{sO} r \qquad (3.30a)$$

which is effectively the same equation as (3.24). Adding the contributions from all the segments we obtain

$$e = [\sum H_i^t T_i \hat{F}_i T_i^t H_i]\, r = \sum F_{iO} r = Fr. \qquad (3.30b)$$

This equation gives the flexibility matrix of the composite member. The load/displacement equations (3.4b) may now be constructed in the usual way. Equation (3.30b) holds for any segmented member, in either two or three dimensions, provided that the T, H and $\hat{F}$ matrices are defined appropriately.

The following sections discuss two practical problems associated with composite members.

3.8. The effect of joints of finite size

A structural joint is not really a mathematical point, any more than a beam is a line. In fact what we have termed a joint is simply an arbitrary point at which we choose to define the end-load and end-displacement of a member.

In real structures the joints are of finite size—usually about 5% of the length of a typical member. If we consider each joint as a rigid body we have the problem of allowing for the rigid regions between the ends of the members and the arbitrary points which represent the joints in the analysis.

It is easy to find the effect of these rigid sections on the $\boldsymbol{K}_{ij}$ matrices of a member. Consider a member AB which has rigid gusset plates attached to its ends, as shown in Fig. 3.12. The points 1 and 2 represent the mathematical joints, and the problem is to find the $\boldsymbol{K}_{ij}$ matrices of the complete member $1AB2$ in terms of the $\boldsymbol{K}_{ij}$ matrices of AB, which are assumed known. The following analysis is presented for a plane member, but the extension to three dimensions follows trivially.

If $\boldsymbol{p}_A$ and $\boldsymbol{p}_B$ are the loads acting on the flexible section of the member at A and B then we have

$$\left.\begin{aligned}\boldsymbol{p}_A &= (\boldsymbol{K}_{11})_{AB}\boldsymbol{d}_A + (\boldsymbol{K}_{12})_{AB}\boldsymbol{d}_B,\\ \boldsymbol{p}_B &= (\boldsymbol{K}_{21})_{AB}\boldsymbol{d}_A + (\boldsymbol{K}_{22})_{AB}\boldsymbol{d}_B.\end{aligned}\right\} \tag{3.31}$$

If these are produced by loads $\boldsymbol{p}_1, \boldsymbol{p}_2$ acting at the points 1, 2 then by statics,

$$\boldsymbol{p}_1 = \begin{bmatrix}1 & 0 & 0\\ 0 & 1 & 0\\ a_y & -a_x & 1\end{bmatrix}\boldsymbol{p}_A, \qquad \boldsymbol{p}_2 = \begin{bmatrix}1 & 0 & 0\\ 0 & 1 & 0\\ b_y & -b_x & 1\end{bmatrix}\boldsymbol{p}_B \tag{3.32a}$$

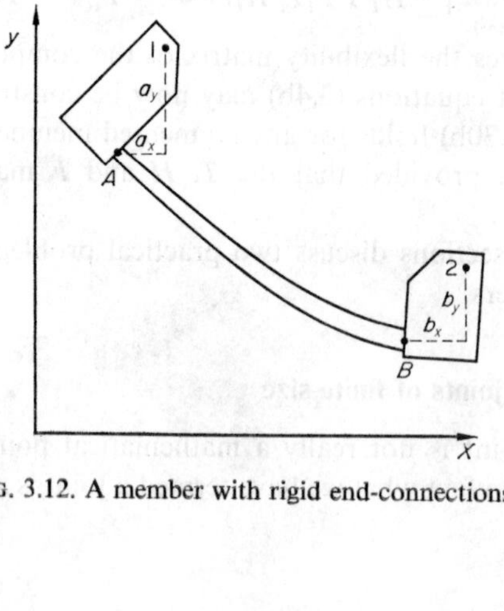

FIG. 3.12. A member with rigid end-connections.

The matrices appearing in these equations are similar in form to the $\boldsymbol{H}$ matrices defined in Section 3.1, except that each one is a function of the *relative* positions of two points of the member rather than of the *absolute* coordinates of one point. We indicate that the matrices are relative rather than absolute by writing them with two subscripts specifying the two points. Thus we write the equations as

$$\boldsymbol{p}_1 = \boldsymbol{H}_{1A}\boldsymbol{p}_A, \quad \boldsymbol{p}_2 = \boldsymbol{H}_{2B}\boldsymbol{p}_B. \tag{3.32b}$$

The displacements are connected by the corresponding equations

$$\boldsymbol{d}_A = \boldsymbol{H}_{1A}{}^t\boldsymbol{d}_1, \; \boldsymbol{d}_B = \boldsymbol{H}_{2B}{}^t\boldsymbol{d}_B.$$

Combining these with (3.31) we obtain

$$\left.\begin{aligned} \boldsymbol{p}_1 &= \boldsymbol{H}_{1A}(\boldsymbol{K}_{11})_{AB}\boldsymbol{H}_{1A}{}^t\boldsymbol{d}_1 + \boldsymbol{H}_{1A}(\boldsymbol{K}_{12})_{AB}\boldsymbol{H}_{2B}{}^t\boldsymbol{d}_2, \\ \boldsymbol{p}_2 &= \boldsymbol{H}_{2B}(\boldsymbol{K}_{21})_{AB}\boldsymbol{H}_{1A}{}^t\boldsymbol{d}_1 + \boldsymbol{H}_{2B}(\boldsymbol{K}_{22})_{AB}\boldsymbol{H}_{2B}{}^t\boldsymbol{d}_2, \end{aligned}\right\} \tag{3.33}$$

which defines the $\boldsymbol{K}_{ij}$ matrices of the complete member. This equation also holds in three dimensions, provided that the $\boldsymbol{K}$ and $\boldsymbol{H}$ matrices are suitably defined. Note that equations (3.33) are still in the member coordinate system associated with the flexible section AB.

3.9. Flexible joint connections

We now consider the case where a member is attached to the joints at its ends by flexible connections, each of which transmits a moment proportional to the difference between the rotation of the end of the member and the rotation of the joint to which it is attached. Such joints occur in bolted frames, and in welded frames after the onset of plasticity. It is true that in both these cases the moment–rotation relation is unlikely to be linear, but we shall show in Chapter 10 that such problems may be solved by iterative means, the structure being treated as a linear system during each iteration. As in the previous section, our aim is to find the stiffness matrices of the complete system of member plus connections, so that in a subsequent analysis we need not consider the rotation which occurs on the member side of each connection.

We consider a straight uniform member of length L and flexural rigidity EI, whose ends 1 and 2 are attached to the joints of a plane frame by connections which exert moments EIk_1/L and EIk_2/L, respectively, per unit difference in rotation. We ignore the length of each connection, so that although the *rotations* may differ, the *translations* of the ends of the member are the same as those of the corresponding joints. The member is shown in Fig. 3.13.

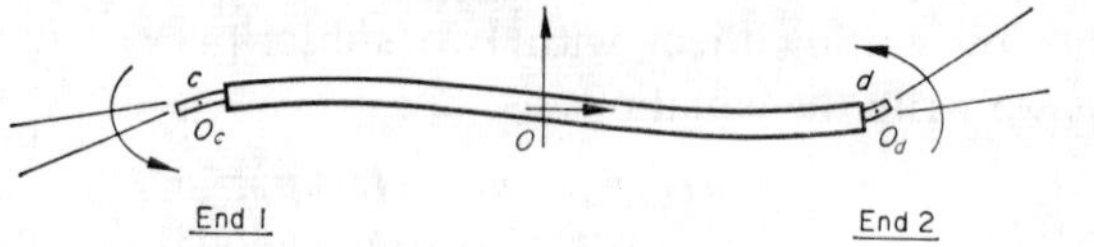

FIG. 3.13. A member with flexible end-connections.

We use expression (3.30b) to find the flexibility matrix of the complete member. If c and d indicate the connections and m indicates the member then from (3.9b) and (3.27b),

$$\boldsymbol{F}_c = \begin{bmatrix} 0 & 0 & 0 \\ 0 & 0 & 0 \\ 0 & 0 & L/EIk_1 \end{bmatrix}, \qquad \boldsymbol{H}_c = \begin{bmatrix} 1 & 0 & 0 \\ 0 & 1 & 0 \\ 0 & L/2 & 1 \end{bmatrix}$$

$$\boldsymbol{F}_d = \begin{bmatrix} 0 & 0 & 0 \\ 0 & 0 & 0 \\ 0 & 0 & L/EIk_2 \end{bmatrix}, \qquad \boldsymbol{H}_d = \begin{bmatrix} 1 & 0 & 0 \\ 0 & 1 & 0 \\ 0 & -L/2 & 1 \end{bmatrix}$$

$$\boldsymbol{F}_m = \begin{bmatrix} L/EA & 0 & 0 \\ 0 & L^3/12EI & 0 \\ 0 & 0 & L/EI \end{bmatrix}$$

and $\boldsymbol{H}_m$ is a unit matrix. No transformation matrices are required, since the coordinate axes for the segments coincide with the coordinate axes for the member. Thus we have

$$\boldsymbol{F} = \boldsymbol{H}_c^t\boldsymbol{F}_c\boldsymbol{H}_c + \boldsymbol{F}_m + \boldsymbol{H}_d^t\boldsymbol{F}_d\boldsymbol{H}_d.$$

Evaluating this expression we obtain

$$\boldsymbol{F} = \begin{bmatrix} L/EA & 0 & 0 \\ 0 & (L^3/12EI)(1 + 3/k_1 + 3/k_2) & (L^2/2EI)(1/k_1 - 1/k_2) \\ 0 & (L^2/2EI)(1/k_1 - 1/k_2) & (L/EI)(1 + 1/k_1 + 1/k_2) \end{bmatrix}$$

which on inversion gives

$$K = \begin{bmatrix} EA/L & 0 & 0 \\ 0 & \dfrac{12EI}{L^3k}(k_1k_2 + k_1 + k_2) & \dfrac{6EI}{L^2k}(k_1 - k_2) \\ 0 & \dfrac{6EI}{L^2k}(k_1 - k_2) & \dfrac{EI}{Lk}(k_1k_2 + 3k_1 + 3k_2) \end{bmatrix}$$

where $k = k_1k_2 + 4(k_1 + k_2) + 12$. The $\boldsymbol{K}_{ij}$ matrices for the complete member now follow in the usual way.

A useful special case of this result occurs when one k value is equal to zero and the other is equal to infinity. This corresponds to a member with a pin joint at one end and a rigid joint at the other

(a) $k_1 = 0, \quad k_2 = \infty$. (Pin joint at end 1, rigid joint at end 2.)

Substituting these values we obtain

$$K = \begin{bmatrix} EA/L & 0 & 0 \\ 0 & 3EI/L^3 & -3EI/2L^2 \\ 0 & -3EI/2L^2 & 3EI/4L \end{bmatrix}$$

From the expression $\boldsymbol{K}_{ij} = \boldsymbol{H}_i\boldsymbol{K}\boldsymbol{H}_j^t$ we obtain

$$K_{11} = \begin{bmatrix} EA/L & 0 & 0 \\ 0 & 3EI/L^3 & 0 \\ 0 & 0 & 0 \end{bmatrix}$$

$$K_{12} = K_{21}^t = \begin{bmatrix} -EA/L & 0 & 0 \\ 0 & -3EI/L^3 & 3EI/L^2 \\ 0 & 0 & 0 \end{bmatrix}$$

$$K_{22} = \begin{bmatrix} EA/L & 0 & 0 \\ 0 & 3EI/L^3 & -3EI/L^2 \\ 0 & -3EI/L^2 & 3EI/L \end{bmatrix}$$

(b) $k_1 = \infty, \quad k_2 = 0$. (Pin joint at end 2, rigid joint at end 1.)

In the same way we have

$$K = \begin{bmatrix} EA/L & 0 & 0 \\ 0 & 3EI/L^3 & 3EI/2L^2 \\ 0 & 3EI/2L^2 & 3EI/4L \end{bmatrix}$$

whence

$$K_{11} = \begin{bmatrix} EA/L & 0 & 0 \\ 0 & 3EI/L^3 & 3EI/L^2 \\ 0 & 3EI/L^2 & 3EI/L \end{bmatrix}$$

$$K_{12} = K_{21}{}^t = \begin{bmatrix} -EA/L & 0 & 0 \\ 0 & -3EI/L^3 & 0 \\ 0 & -3EI/L^2 & 0 \end{bmatrix}$$

$$K_{22} = \begin{bmatrix} EA/L & 0 & 0 \\ 0 & 3EI/L^3 & 0 \\ 0 & 0 & 0 \end{bmatrix}$$

The reader may notice that the last term of the matrix K_{22} in case (a) (and K_{11} in case (b)) has been reduced from $4EI/L$ to $3EI/L$ by the presence of the pin joint. This corresponds to the modification of rotational stiffness normally made in the moment distribution process when a member has a pin joint at one end. Finally it may be noted that if k_1 and k_2 are both zero then all the flexural terms in the stiffness matrices become zero, which corresponds to the fact that a member which is pinned at both ends may be treated as a member of zero flexural rigidity rigidly attached to the joints at its ends.

3.10. The calculation of equivalent joint loads

We now return to the problem of calculating equivalent joint loads. These may be due to external loads applied along the length of a member or they may arise from initial strains (i.e. temperature changes or erection misalignments). We begin by considering the problem of external loads.

As discussed in Section 1.4, we can regard the loading on a structure as the sum of

(a) The external loading, plus a set of concentrated joint loads $-\mathbf{p}_{\text{equiv}}$ whose components are of such magnitudes as to prevent any displacement of the joints.

(b) A set of loads $\mathbf{p}_{\text{equiv}}$ acting at the joints of the structure. These are the loads used in the analysis of the complete structure.

Thus our problem is to find the end-loads for a general member rigidly held at its ends and carrying an arbitrary distribution of loading along its length.

In Section 3.6 we found that a simple spring analogue suggested a fruitful line of attack, even though it did not illustrate all aspects of the general problem. We therefore preface our analysis by a brief look at the two springs shown in Fig. 3.14. In this example our natural approach is to say that the two springs are "in parallel", so that the displacement of the loaded point is given by $d = p/(k_1 + k_2)$. It follows that $p_1 = k_1 p/(k_1 + k_2)$ and $p_2 = k_2 p/(k_1 + k_2)$. These may be written in the form $p_1 = kf_2 p$, $p_2 = kf_1 p$ where f_1, f_2 are the flexibilities of the individual springs and $k = k_1 k_2/(k_1 + k_2)$ is the overall stiffness of the springs when connected "in series".

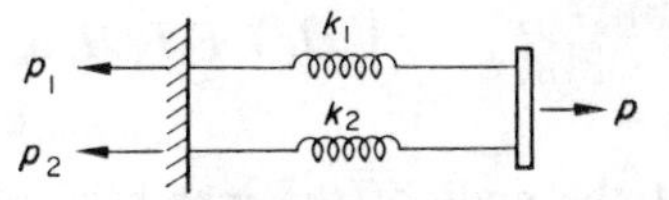

FIG. 3.14. Two springs in parallel.

We now return to the general problem. We consider the plane member shown in Fig. 3.15, loaded by a single concentrated load $\mathbf{w}_S$ (the term "load" may here include a moment component) at an arbitrary point S. The member is kept in equilibrium by loads $-\mathbf{q}_1$, $\mathbf{q}_2$ applied to the arms at O.

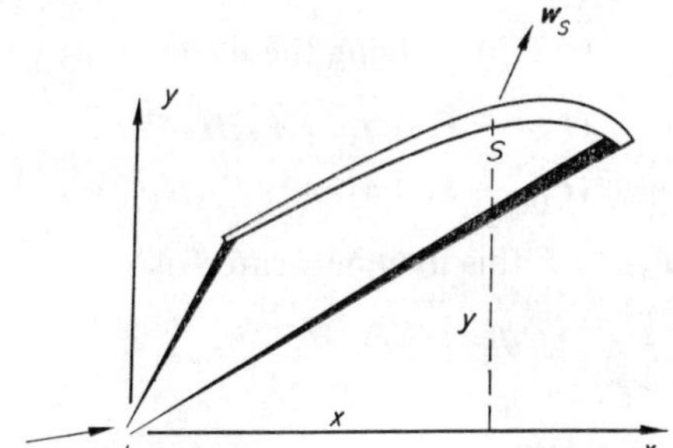

FIG. 3.15. A general member carrying a point load.

From the analysis of Section 3.6 it follows that if the section from the point S to end 2 were made rigid the relative displacement of the ends of the two arms would be

$$\boldsymbol{e}^{(1)} = \boldsymbol{F}_{1S}\boldsymbol{q}_1$$

where $\boldsymbol{F}_{1S} = \int_0^S \boldsymbol{H}_s^t \boldsymbol{T}_s \hat{\boldsymbol{F}}_s \boldsymbol{T}_s^t \boldsymbol{H}_s \, ds$, the flexibility matrix of the section of member from end 1 to the point S. Similarly if the section from end 1 to the point S were rigid the relative displacement would be

$$\boldsymbol{e}^{(2)} = \boldsymbol{F}_{S2}\boldsymbol{q}_2$$

where

$$\boldsymbol{F}_{S2} = \int_S^L \boldsymbol{H}_s^t \boldsymbol{T}_s \hat{\boldsymbol{F}}_s \boldsymbol{T}_s^t \boldsymbol{H}_s \, ds.$$

The condition that the ends of the member are rigidly fixed implies that there is no relative displacement of the ends of the two arms, i.e. that $\boldsymbol{e}^{(1)} + \boldsymbol{e}^{(2)} = \boldsymbol{0}$. Thus we have

$$\boldsymbol{F}_{1S}\boldsymbol{q}_1 + \boldsymbol{F}_{S2}\boldsymbol{q}_2 = \boldsymbol{0}. \tag{3.34}$$

Since the member is in equilibrium under the loads $\boldsymbol{w}_S$, $-\boldsymbol{q}_1$, $\boldsymbol{q}_2$ we have

$$\boldsymbol{H}_S(\boldsymbol{q}_2 - \boldsymbol{q}_1) + \boldsymbol{w}_S = \boldsymbol{0} \tag{3.35}$$

where $\boldsymbol{H}_S$ is given by (3.21b). Solving these equations we obtain

$$(\boldsymbol{F}_{1S} + \boldsymbol{F}_{S2})\,\boldsymbol{q}_1 = \boldsymbol{F}_{S2}\boldsymbol{H}_S^{-1}\boldsymbol{w}_S,$$
$$(\boldsymbol{F}_{1S} + \boldsymbol{F}_{S2})\,\boldsymbol{q}_2 = -\boldsymbol{F}_{1S}\boldsymbol{H}_S^{-1}\boldsymbol{w}_S,$$

and since $\boldsymbol{F}_{1S} + \boldsymbol{F}_{S2} = \boldsymbol{F}$ this may be written as

$$\boldsymbol{q}_1 = \boldsymbol{K}\boldsymbol{F}_{S2}\boldsymbol{H}_S^{-1}\boldsymbol{w}_S,$$
$$\boldsymbol{q}_2 = -\boldsymbol{K}\boldsymbol{F}_{1S}\boldsymbol{H}_S^{-1}\boldsymbol{w}_S.$$

At this stage the similarity to the parallel-spring problem is obvious. The corresponding end -loads $\boldsymbol{p}_1$, $\boldsymbol{p}_2$ are given by the equations $\boldsymbol{p}_1 = \boldsymbol{H}_1\boldsymbol{q}_1$, $\boldsymbol{p}_2 = \boldsymbol{H}_2\boldsymbol{q}_2$, where $\boldsymbol{H}_1$ and $\boldsymbol{H}_2$ are defined by (3.6).

Finally we have

$$\left.\begin{aligned}(\boldsymbol{p}_{\text{equiv}})_1 &= -\boldsymbol{p}_1 = -\boldsymbol{H}_1\boldsymbol{K}\boldsymbol{F}_{S2}\boldsymbol{H}_S^{-1}\boldsymbol{w}_S,\\ (\boldsymbol{p}_{\text{equiv}})_2 &= -\boldsymbol{p}_2 = \quad \boldsymbol{H}_2\boldsymbol{K}\boldsymbol{F}_{1S}\boldsymbol{H}_S^{-1}\boldsymbol{w}_S.\end{aligned}\right\} \tag{3.36}$$

Note that $\boldsymbol{H}_S^{-1}$ is equal to $\boldsymbol{H}_S$ with the signs of the off-diagonal elements reversed.

Expressions for the end-loads equivalent to a general load distribution may be obtained by integrating equations (3.36). If $\boldsymbol{w}_S$ is now regarded as the load per unit length at point S we have

$$\left.\begin{aligned}(\boldsymbol{p}_{\text{equiv}})_1 &= -\boldsymbol{H}_1\boldsymbol{K}\int_0^L \boldsymbol{F}_{S2}\boldsymbol{H}_S^{-1}\boldsymbol{w}_S\,dS,\\ (\boldsymbol{p}_{\text{equiv}})_2 &= \quad \boldsymbol{H}_2\boldsymbol{K}\int_0^L \boldsymbol{F}_{1S}\boldsymbol{H}_S^{-1}\boldsymbol{w}_S\,dS.\end{aligned}\right\} \tag{3.37}$$

The effect of initial strains may be dealt with by a very similar analysis. As in Section 3.6, we first consider the effect of strain in one isolated segment of length ds, as shown in Fig. 3.16. Let the axial strain be $\hat{\epsilon}_x$ and

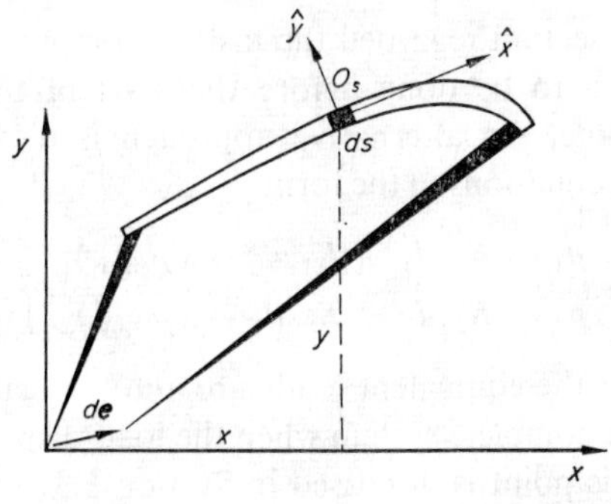

FIG. 3.16. Effect of strain in a segment of length *ds*.

the rotational strain (i.e. the change in curvature per unit length) be $\hat{\epsilon}_\theta$. As previously, we assume that there is no shear strain. Thus the deformation vector for the unstressed segment, expressed in the local coordinate system $\hat{x}$, $\hat{y}$, is

$$\begin{bmatrix} \hat{\epsilon}_x \\ 0 \\ \hat{\epsilon}_\theta \end{bmatrix} ds = \hat{\boldsymbol{e}}_s \, ds,$$

and the corresponding relative displacement of the two arms at O is given by (3.23) as $d\boldsymbol{e} = \boldsymbol{H}_s^t \boldsymbol{T}_s \hat{\boldsymbol{e}}_s \, ds$. Hence the total relative displacement at O due to initial strains in the complete unstressed member is

$$\boldsymbol{e} = \int_0^L \boldsymbol{H}_s^t \boldsymbol{T}_s \hat{\boldsymbol{e}}_s \, ds. \tag{3.38}$$

We may reduce this relative displacement to zero by applying a self-equilibrating load pair $-\boldsymbol{r}$, $\boldsymbol{r}$ to the ends of the two arms, where $\boldsymbol{r} = -\boldsymbol{K}\boldsymbol{e}$. The corresponding end-loads $\boldsymbol{p}_1$, $\boldsymbol{p}_2$ follow in the usual way from (3.6b) so that finally we have

$$\left.\begin{aligned} (\boldsymbol{p}_{\text{equiv}})_1 &= -\boldsymbol{p}_1 = \boldsymbol{H}_1 \boldsymbol{K} \int_0^L \boldsymbol{H}_s^t \boldsymbol{T}_s \hat{\boldsymbol{e}}_s \, ds, \\ (\boldsymbol{p}_{\text{equiv}})_2 &= -\boldsymbol{p}_2 = \boldsymbol{H}_2 \boldsymbol{K} \int_0^L \boldsymbol{H}_s^t \boldsymbol{T}_s \hat{\boldsymbol{e}}_s \, ds. \end{aligned}\right\} \tag{3.39}$$

We have in this section regarded the task of computing equivalent joint loads as something to be done before the start of the analysis of the structure as a whole. An alternative approach is to write the member load/displacement equations in the form

$$\left.\begin{aligned} \boldsymbol{p}_1 &= \boldsymbol{K}_{11}\boldsymbol{d}_1 + \boldsymbol{K}_{12}\boldsymbol{d}_2 - (\boldsymbol{p}_{\text{equiv}})_1 \\ \boldsymbol{p}_2 &= \boldsymbol{K}_{21}\boldsymbol{d}_1 + \boldsymbol{K}_{22}\boldsymbol{d}_2 - (\boldsymbol{p}_{\text{equiv}})_2. \end{aligned}\right\} \tag{3.40}$$

If this is done then the equivalent loads are automatically included in the load vector for the complete system when the load/displacement equations are assembled. This point is discussed in Section 5.3.

3.11. The replacement of distributed mass

When a structure vibrates, the mass of the members gives rise to a distributed inertia loading. Any analytical procedure based on joint displacements must replace this distributed loading by some form of equivalent loading at the joints.

Two lines of attack are available. The first, which has a direct intuitive appeal, consists of replacing the distributed mass of the structure by a set of concentrated masses situated at the joints. In its simplest form this merely involves replacing the mass of each member by two statically equivalent concentrated masses at the two ends. Thus a uniform beam of mass M contributes masses $M/2$ to the two joints at its ends. These masses give rise to inertia loads at the joints

$$\begin{bmatrix} p_x \\ p_y \\ m \end{bmatrix}_{1,2} = -\begin{bmatrix} M/2 & 0 & 0 \\ 0 & M/2 & 0 \\ 0 & 0 & 0 \end{bmatrix} \begin{bmatrix} \ddot{\delta}_x \\ \ddot{\delta}_y \\ \ddot{\theta} \end{bmatrix}_{1,2} \tag{3.41a}$$

which may be written as

$$(\boldsymbol{p}_{\text{equiv}})_i = -\boldsymbol{M}\ddot{\boldsymbol{d}}_i, \qquad i = 1,2. \tag{3.41b}$$

The second approach makes use of the results of the previous section. It is apparent that equations (3.37) will give us the equivalent loads we require, provided we can calculate the distributed inertia loading $\boldsymbol{w}_S$. However, this loading is a function of the local displacement at each point of the member, and an exact calculation of this displacement involves solving the partial differential equations of motion of the structure.

We may avoid this difficulty by assuming that the displaced form of the member is what it would be under *static* conditions, with the same end-displacements and no loading except at the ends. This is clearly only approximately true, but it is generally a closer approximation to the truth than the simple mass-replacement process described above. This leads us to the problem of finding the displacement $\boldsymbol{u}_S$ at an arbitrary point of a member with given end-displacements $\boldsymbol{d}_1$, $\boldsymbol{d}_2$.

This displacement may be derived using Betti's reciprocal theorem. Consider the two load and displacement systems shown in Fig. 3.17. In System 1 the end displacements are zero, while in System 2 there is no load at the point S. Betti's theorem states that

$$-(\boldsymbol{p}^*_{\text{equiv}})_1{}^t\boldsymbol{d}_1 - (\boldsymbol{p}^*_{\text{equiv}})_2{}^t\boldsymbol{d}_2 + \boldsymbol{w}_S{}^{*t}\boldsymbol{u}_S = \boldsymbol{p}_1{}^t\boldsymbol{d}_1{}^* + \boldsymbol{p}_2{}^t\boldsymbol{d}_2{}^* + \boldsymbol{w}_S{}^t\boldsymbol{u}_S{}^*.$$

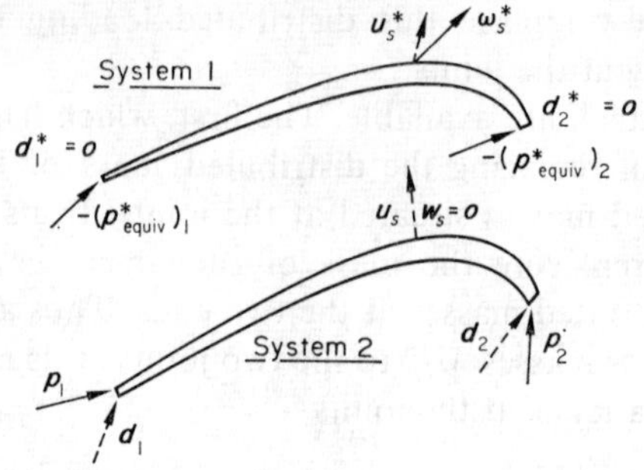

FIG. 3.17. A pair of load and displacement systems.

Since $\boldsymbol{d}_1{}^*$, $\boldsymbol{d}_2{}^*$ and $\boldsymbol{w}_S$ are all zero it follows that

$$\boldsymbol{w}_S{}^{*t}\boldsymbol{u}_S = (\boldsymbol{p}^*_{\text{equiv}})_1{}^t\boldsymbol{d}_1 + (\boldsymbol{p}^*_{\text{equiv}})_2{}^t\boldsymbol{d}_2. \tag{3.42}$$

Substituting from (3.36) for $(\boldsymbol{p}^*_{\text{equiv}})_1$ and $(\boldsymbol{p}^*_{\text{equiv}})_2$ in (3.42) and "cancelling" the common multiplier $\boldsymbol{w}_S{}^{*t}$ we obtain

$$\boldsymbol{u}_S = -(\boldsymbol{H}_S{}^{-1})^t\boldsymbol{F}_{S2}\boldsymbol{K}\boldsymbol{H}_1{}^t\boldsymbol{d}_1 + (\boldsymbol{H}_S{}^{-1})^t\boldsymbol{F}_{1S}\boldsymbol{K}\boldsymbol{H}_2{}^t\boldsymbol{d}_2. \tag{3.43}$$

If the line element at S has mass ρ_S per unit length, the inertia loading per unit length is

$$\boldsymbol{w}_S = -\begin{bmatrix} \rho_S & 0 & 0 \\ 0 & \rho_S & 0 \\ 0 & 0 & 0 \end{bmatrix} \ddot{\boldsymbol{u}}_S = -\boldsymbol{M}_S\ddot{\boldsymbol{u}}_S$$

(Note that in general $\boldsymbol{w}_S$ and $\boldsymbol{u}_S$ have moment and rotation components respectively. We are here ignoring the local rotational inertia of an infinitesimal element.) Substituting for $\boldsymbol{w}_S$ in (3.37) gives

$$\begin{aligned} (\boldsymbol{p}_{\text{equiv}})_1 &= -\boldsymbol{H}_1\boldsymbol{K}\left[\int_0^L \boldsymbol{F}_{S2}\boldsymbol{M}_{SO}\boldsymbol{F}_{S2}\,dS\right]\boldsymbol{K}\boldsymbol{H}_1{}^t\ddot{\boldsymbol{d}}_1 \\ &\qquad + \boldsymbol{H}_1\boldsymbol{K}\left[\int_0^L \boldsymbol{F}_{S2}\boldsymbol{M}_{SO}\boldsymbol{F}_{1S}\,dS\right]\boldsymbol{K}\boldsymbol{H}_2{}^t\ddot{\boldsymbol{d}}_2, \\ (\boldsymbol{p}_{\text{equiv}})_2 &= \boldsymbol{H}_2\boldsymbol{K}\left[\int_0^L \boldsymbol{F}_{1S}\boldsymbol{M}_{SO}\boldsymbol{F}_{S2}\,dS\right]\boldsymbol{K}\boldsymbol{H}_1{}^t\ddot{\boldsymbol{d}}_1 \\ &\qquad - \boldsymbol{H}_2\boldsymbol{K}\left[\int_0^L \boldsymbol{F}_{1S}\boldsymbol{M}_{SO}\boldsymbol{F}_{1S}\,dS\right]\boldsymbol{K}\boldsymbol{H}_2{}^t\ddot{\boldsymbol{d}}_2 \end{aligned} \tag{3.44a}$$

where $M_{SO} = H_S^{-1} M_S (H_S^{-1})^t$. The matrix M_{SO} is equal to

$$\rho_S \begin{bmatrix} 1 & 0 & -y \\ 0 & 1 & x \\ -y & x & (x^2 + y^2) \end{bmatrix}$$

and can be thought of as representing the inertia of a unit length of beam at S, as measured at the origin O. We may write (3.44a) in a form similar to (3.41b),

$$(\boldsymbol{p}_{\text{equiv}})_i = -\boldsymbol{M}_{ij}\ddot{\boldsymbol{d}}_j, \quad i,j = 1,2. \tag{3.44b}$$

The matrices $\boldsymbol{M}_{ij}$ appearing in (3.44b) are referred to as *consistent mass matrices.* Note the important difference between (3.41b) and (3.44b). In the second of the two approaches each equivalent end-load depends on both $\ddot{\boldsymbol{d}}_1$ and $\ddot{\boldsymbol{d}}_2$. One cannot, therefore, visualize the result as a set of physical masses or inertias placed at the joints of a weightless structure.

The mass matrix $\boldsymbol{M}$ defined by (3.41) is independent of the coordinate axes chosen for the member. The consistent mass matrices $\boldsymbol{M}_{ij}$ defined by (3.44) transform into global coordinates in exactly the same way as the member $\boldsymbol{K}_{ij}$ matrices.

Evaluation of the matrices $\boldsymbol{M}_{ij}$ from (3.44a) is tedious but not difficult in principle. For a straight uniform beam of total mass M the matrices are

$$\boldsymbol{M}_{11} = \frac{M}{420}\begin{bmatrix} 140 & 0 & 0 \\ 0 & 156 & 22L \\ 0 & 22L & 4L^2 \end{bmatrix}$$

$$\boldsymbol{M}_{12} = \boldsymbol{M}_{21}{}^t = \frac{M}{420}\begin{bmatrix} 70 & 0 & 0 \\ 0 & 54 & -13L \\ 0 & 13L & -3L^2 \end{bmatrix}$$

$$\boldsymbol{M}_{22} = \frac{M}{420}\begin{bmatrix} 140 & 0 & 0 \\ 0 & 156 & -22L \\ 0 & -22L & 4L^2 \end{bmatrix}$$

These results were first derived by Archer (1963) and independently by Leckie and Lindberg (1963). Further refinements of the analysis may be found in Przemieniecki (1968). In the case of curved or non-uniform members the integrals defining $\boldsymbol{F}_{1S}$ and $\boldsymbol{F}_{S2}$ and the integrals appearing in (3.44a) will usually have to be evaluated numerically.

The use of the mass matrices derived in this section and the relative accuracy of the two approaches are discussed in Section 5.5.

CHAPTER 4

The Elastic Properties of Single Elements

(b) *An Introduction to Area and Volume Elements*

An attempt to treat area and volume elements from the same viewpoint as the line elements discussed in the previous chapter reveals fundamental differences between the two types of element. These differences are:

1. In a line element forming part of a skeletal structure the only boundaries on which compatibility of displacement must be satisfied are the end cross-sections. The "plane sections remain plane" assumption of simple bending theory automatically makes any line element conformable. An area or volume element, on the other hand, usually has to conform on all its boundaries, and this restriction severely limits the type of deformation which can be permitted.
2. The stresses and stress-resultants at an interior point of a line element can be expressed in terms of the nodal (i.e. joint) loads from statics alone, without any consideration of deformations (provided the latter are small). In a continuum element this is not possible.

Although these differences have a considerable influence on the analysis, the present chapter is laid out in much the same way as the previous one. We defer the calculation of equivalent nodal loads and inertias to the end of the chapter, and begin by considering the simplest element used in continuum analysis—the plane triangle. For the sake of simplicity most of the chapter is concerned with plane stress problems. For the same reason the detailed analysis is presented in ordinary rectangular coordinates, rather than in the "natural" or "area" coordinates developed by Argyris, Zienkiewicz and others. The reader whose primary interest is in continuum problems should regard the present treatment merely as an introduction to a standard text such as Zienkiewicz (1971).

The analysis presented in this chapter is based on the vectors $\boldsymbol{r}$ and $\boldsymbol{e}$ introduced in Chapter 2. Although these vectors have less direct physical meaning than is the case with line elements they still provide a convenient tool in the construction of the nodal load/displacement equations.

4.1. The basic equations of plane stress

For convenience we summarize here the standard relationships of elastic plane-stress theory.

We consider a sheet of isotropic material of thickness h lying in the xy plane. We imagine that all loads applied to the material are also in the xy plane, so that the only non-zero stress components are σ_x, σ_y, τ_{xy}, as shown in Fig. 4.1. At any point (x,y) in the material we can define four vectors, all of which may be functions of x and y. The vectors are:

The displacement $\boldsymbol{u} = \begin{bmatrix} u_x \\ u_y \end{bmatrix}$. The body force per unit volume $\boldsymbol{w} = \begin{bmatrix} w_x \\ w_y \end{bmatrix}$.

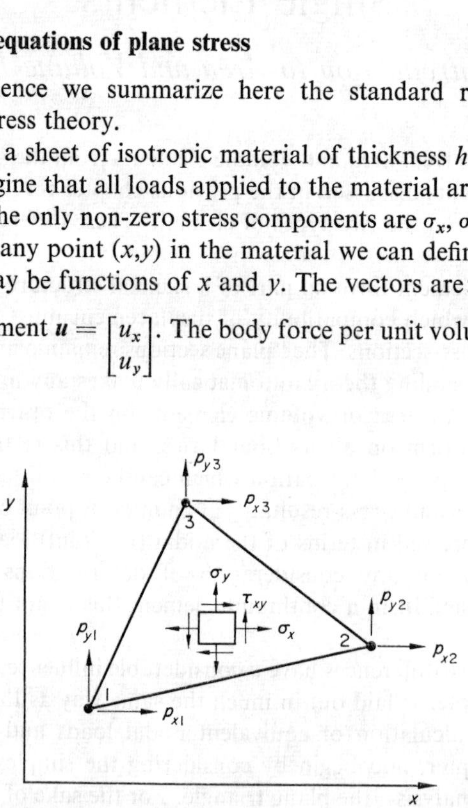

FIG. 4.1. Stresses and nodal forces in a simple triangular element for plane stress.

The strain $\boldsymbol{\epsilon} = \begin{bmatrix} \epsilon_x \\ \epsilon_y \\ \gamma_{xy} \end{bmatrix}$. The stress $\boldsymbol{\sigma} = \begin{bmatrix} \sigma_x \\ \sigma_y \\ \tau_{xy} \end{bmatrix}$.

The vectors $\boldsymbol{\epsilon}$ and $\boldsymbol{\sigma}$ are related by the equations of elasticity

$$\boldsymbol{\sigma} = \boldsymbol{D\epsilon}, \qquad \boldsymbol{\epsilon} = \boldsymbol{D}^{-1}\boldsymbol{\sigma} \tag{4.1}$$

where

$$\boldsymbol{D} = \frac{E}{1-\nu^2}\begin{bmatrix} 1 & \nu & 0 \\ \nu & 1 & 0 \\ 0 & 0 & \dfrac{(1-\nu)}{2} \end{bmatrix}, \quad \boldsymbol{D}^{-1} = \frac{1}{E}\begin{bmatrix} 1 & -\nu & 0 \\ -\nu & 1 & 0 \\ 0 & 0 & 2(1+\nu) \end{bmatrix} \tag{4.2}$$

From the elementary definitions of direct and shear strains we have

$$\begin{bmatrix} \epsilon_x \\ \epsilon_y \\ \gamma_{xy} \end{bmatrix} = \begin{bmatrix} \partial/\partial x & 0 \\ 0 & \partial/\partial y \\ \partial/\partial y & \partial/\partial x \end{bmatrix} \begin{bmatrix} u_x \\ u_y \end{bmatrix} \tag{4.3a}$$

which may be written as

$$\boldsymbol{\epsilon} = \boldsymbol{\partial u}. \tag{4.3b}$$

(This equation implies that the three components of strain are not independent functions.)

By considering the equilibrium of a small element of material we obtain

$$\begin{bmatrix} \partial/\partial x & 0 & \partial/\partial y \\ 0 & \partial/\partial y & \partial/\partial x \end{bmatrix} \begin{bmatrix} \sigma_x \\ \sigma_y \\ \tau_{xy} \end{bmatrix} + \begin{bmatrix} w_x \\ w_y \end{bmatrix} = \begin{bmatrix} 0 \\ 0 \end{bmatrix} \tag{4.4a}$$

which may be written

$$\boldsymbol{\partial}^t\boldsymbol{\sigma} + \boldsymbol{w} = \boldsymbol{0}. \tag{4.4b}$$

Once again we see how matrix notation brings out the relationship between the differential operators in equations (4.3) and (4.4).

4.2. The simple triangular element

This is the simplest element for plane stress analysis and is shown in Fig. 4.1. As mentioned earlier, we begin by assuming that there are no body forces on the element material, i.e. that $\boldsymbol{w} = \boldsymbol{0}$.

In Section 1.3 we argued that it is desirable to choose the deformation mode of an element in such a way that the element is conformable. However, in the case of a simple triangular element there is no real choice. For since there are three nodes, the vectors $\boldsymbol{p}$ and $\boldsymbol{d}$ each have six components. Consequently the vectors $\boldsymbol{r}$ and $\boldsymbol{e}$ each have three components, and since there are three components in the strain vector $\boldsymbol{\epsilon}$ we only have enough parameters in $\boldsymbol{e}$ to define a constant strain system.

At first sight such an element seems a very crude approximation for use in situations where stresses are known to vary continuously. However, since strains are constant the straight line boundaries of the element remain straight after deformation. The element is therefore conformable. The conditions of equilibrium (4.4) are also satisfied at all points within the element. This, as we shall see later, is not always true in more complex elements.

The simple form of the strain system allows us to give direct physical significance to the vectors $\boldsymbol{e}$ and $\boldsymbol{r}$. If we make $\boldsymbol{e}$ identical with the (constant) strain vector $\boldsymbol{\epsilon}$ then the corresponding components of $\boldsymbol{r}$ are the stresses $\boldsymbol{\sigma}$ multiplied by the volume V of the element (the volume is introduced here to satisfy the condition $\boldsymbol{r}^t\boldsymbol{e}^* = V(\sigma_x\epsilon_x{}^* + \sigma_y\epsilon_y{}^* + \tau_{xy}\gamma_{xy}{}^*)$ = virtual work done on the element during a virtual deformation $\boldsymbol{e}^*$). Thus

$$\boldsymbol{r} = V\begin{bmatrix}\sigma_x\\ \sigma_y\\ \tau_{xy}\end{bmatrix}, \qquad \boldsymbol{e} = \begin{bmatrix}\epsilon_x\\ \epsilon_y\\ \gamma_{xy}\end{bmatrix}$$

The vectors $\boldsymbol{r}$ and $\boldsymbol{e}$ are related by the usual equations

$$\boldsymbol{r} = \boldsymbol{K}\boldsymbol{e}, \quad \boldsymbol{e} = \boldsymbol{F}\boldsymbol{r} \tag{4.5a}$$

where from (4.2) we have

$$K = \frac{EV}{1-\nu^2}\begin{bmatrix} 1 & \nu & 0 \\ \nu & 1 & 0 \\ 0 & 0 & \dfrac{(1-\nu)}{2} \end{bmatrix}, \quad F = \frac{1}{EV}\begin{bmatrix} 1 & -\nu & 0 \\ -\nu & 1 & 0 \\ 0 & 0 & 2(1+\nu) \end{bmatrix} \tag{4.5b}$$

Note that $\boldsymbol{K}$ and $\boldsymbol{F}$ may be made purely diagonal by choosing the first two components of $\boldsymbol{e}$ to be the "hydrostatic" strain $(\epsilon_x + \epsilon_y)/2$ and the "deviatoric" strain $(\epsilon_x - \epsilon_y)/2$.

The next step is to find the equilibrium equations which express the nodal loads $\boldsymbol{p}$ in terms of the vector $\boldsymbol{r}$. This may be done by taking each side of the triangle in turn and replacing the distributed boundary stresses by equivalent forces at the nodes. For example, the stresses on the boundary 1 . . . 2 in Fig. 4.1 sum to a stress-resultant which has a component

$$h[\sigma_x(y_2 - y_1) - \tau_{xy}(x_2 - x_1)]$$

in the direction of the x-axis. Similarly the stresses on 1 . . . 3 produce a stress-resultant with a component

$$h[-\sigma_x(y_3 - y_1) + \tau_{xy}(x_3 - x_1)]$$

in the same direction. It follows that

$$\begin{aligned} p_{x1} &= \frac{h}{2}[\sigma_x(y_2 - y_1) - \tau_{xy}(x_2 - x_1)] + \frac{h}{2}{-\sigma_x}(y_3 - y_1) + \tau_{xy}(x_3 - x_1)] \\ &= \frac{h}{2}[\sigma_x(y_2 - y_3) - \tau_{yx}(x_2 - x_3)]. \end{aligned}$$

Similar reasoning applied to the other elements of $\boldsymbol{p}$ gives

$$\boldsymbol{p} = \begin{bmatrix} p_{x1} \\ p_{y1} \\ \hline p_{x2} \\ p_{y2} \\ \hline p_{x3} \\ p_{y3} \end{bmatrix} = \frac{h}{2V}\begin{bmatrix} y_2 - y_3 & 0 & x_3 - x_2 \\ 0 & x_3 - x_2 & y_2 - y_3 \\ \hline y_3 - y_1 & 0 & x_1 - x_3 \\ 0 & x_1 - x_3 & y_3 - y_1 \\ \hline y_1 - y_2 & 0 & x_2 - x_1 \\ 0 & x_2 - x_1 & y_1 - y_2 \end{bmatrix} \boldsymbol{r} \tag{4.6a}$$

Equation (4.6a) gives the form of the equilibrium equation $\boldsymbol{p} = \boldsymbol{Hr}$ for the constant-strain triangular element. We may write this in the form

$$\left.\begin{aligned} \boldsymbol{p}_1 &= \boldsymbol{H}_1\boldsymbol{r}, \\ \boldsymbol{p}_2 &= \boldsymbol{H}_2\boldsymbol{r}, \\ \boldsymbol{p}_3 &= \boldsymbol{H}_3\boldsymbol{r}. \end{aligned}\right\} \tag{4.6b}$$

From the general virtual-work argument presented in Section 2.3 we should expect the corresponding relation between $\boldsymbol{e}$ and the nodal displacements $\boldsymbol{d}$ to be $\boldsymbol{e} = \boldsymbol{H}^t\boldsymbol{d}$, i.e.

$$\begin{bmatrix} \epsilon_x \\ \epsilon_y \\ \gamma_{xy} \end{bmatrix} = \frac{h}{2V} \left[\begin{array}{cc|cc|cc} y_2 - y_3 & 0 & y_3 - y_1 & 0 & y_1 - y_2 & 0 \\ \cdot & \cdot & \cdot & \cdot & \cdot & \cdot \\ \cdot & \cdot & \cdot & \cdot & \cdot & \cdot \end{array}\right] \begin{bmatrix} \delta_{x1} \\ \delta_{y1} \\ \delta_{x2} \\ \delta_{y2} \\ \delta_{x3} \\ \delta_{y3} \end{bmatrix} \tag{4.7}$$

It is easy to verify that this is the case. Consider a distortion of the triangle defined by displacements δ_{x1}, δ_{x2}, δ_{x3}, as shown in Fig. 4.2. Clearly

$$\epsilon_x = \frac{\delta_{x2} - \delta_{x4}}{x_2 - x_4} = \left[\delta_{x2} - \frac{\delta_{x3}(y_2 - y_1) + \delta_{x1}(y_3 - y_2)}{y_3 - y_1}\right] \Big/ (x_2 - x_4)$$

$$= \frac{\delta_{x1}(y_2 - y_3) + \delta_{x2}(y_3 - y_1) + \delta_{x3}(y_1 - y_2)}{2 \times \text{area of triangle}}$$

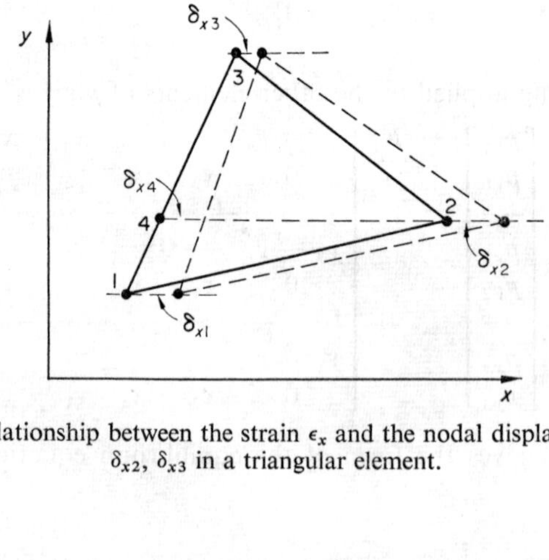

FIG. 4.2. Relationship between the strain ϵ_x and the nodal displacements δ_{x1}, δ_{x2}, δ_{x3} in a triangular element.

which corresponds to the first of the three scalar equations in (4.7). The other equations may be verified in a similar manner. Equation (4.7) may be written in the usual way as

$$\boldsymbol{e} = \boldsymbol{H}_1{}^t\boldsymbol{d}_1 + \boldsymbol{H}_2{}^t\boldsymbol{d}_2 + \boldsymbol{H}_3{}^t\boldsymbol{d}_3. \tag{4.8}$$

Combining (4.5a), (4.6b) and (4.8) we obtain the nodal load/displacement equations $\boldsymbol{p}_i = \boldsymbol{H}_i\boldsymbol{K}\boldsymbol{H}_j{}^t\boldsymbol{d}_j$, $(i, j = 1,2,3)$. The evaluation of the nodal stiffness matrices $\boldsymbol{K}_{ij} = \boldsymbol{H}_i\boldsymbol{K}\boldsymbol{H}_j{}^t$ is rather a tedious exercise in matrix multiplication, which can well be left to a computer.

4.3. The simple rectangular element

The simple triangle of the previous section has been used in many successful applications of the finite element method. However, the fact that all three stress components are constant within each element means that very small elements have to be used in regions of rapid stress variation.

If we are to allow strains to vary within an element we clearly need some extra parameters in the vector $\boldsymbol{e}$. We may obtain these by introducing an extra node, i.e. by changing to a quadrilateral element. With four nodes we have eight components in the vectors $\boldsymbol{p}$ and $\boldsymbol{d}$, so that there are five components in each of the vectors $\boldsymbol{r}$ and $\boldsymbol{e}$. Thus the extra node allows us to insert two additional parameters in the element deformation vector. In this section we consider the rectangle shown in Fig. 4.3, rather than a general quadrilateral. This simplifies the algebra and involves no real loss of generality, since any quadrilateral may be mapped onto a rectangle by a suitable coordinate transformation. As before, we assume that the element has constant thickness h.

The simplicity of the triangular element allowed us to give a direct physical meaning to the components of the vectors $\boldsymbol{r}$ and $\boldsymbol{e}$, and to derive the $\boldsymbol{K}$ and $\boldsymbol{H}$ matrices by a physical argument. Although we could continue with this approach in the case of the rectangular element, we shall in fact take this opportunity to present a more general method which is important in the development of complex elements. In this approach the vectors $\boldsymbol{r}$ and $\boldsymbol{e}$ keep their logical importance as defining the stress and deformation state of the element, but their physical interpretation becomes less direct.

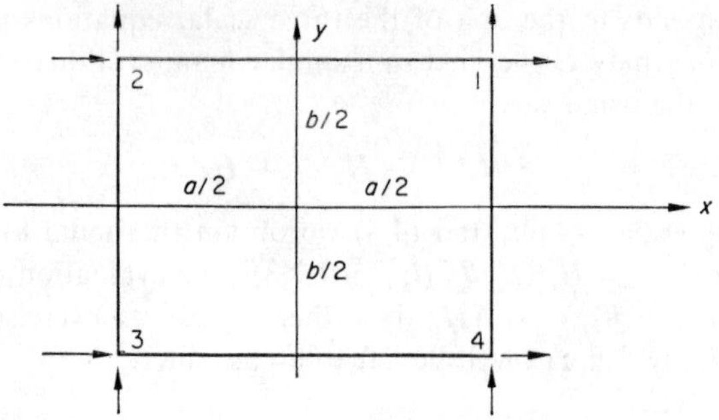

FIG. 4.3. A simple rectangular element for plane stress.

The two extra parameters in the element deformation vector allow us to add two additional modes of deformation to the constant strains which were the only possible deformations for the triangular element. Although at first sight there appears to be considerable freedom of choice, there are in fact only two varying strain functions for which straight-line boundaries remain straight after deformation (the condition for the element to be conformable). These are the ones in which ϵ_x varies linearly with y and ϵ_y varies linearly with x. We therefore write ϵ_x and ϵ_y in the form

$$\left.\begin{aligned} \epsilon_x &= \frac{\partial u_x}{\partial x} = e_1 + e_4 \frac{2y}{b}, \\ \epsilon_y &= \frac{\partial u_y}{\partial y} = e_2 + e_5 \frac{2x}{a}. \end{aligned}\right\} \tag{4.9}$$

Integrating these two equations we obtain the displacements as the bilinear functions

$$\left.\begin{aligned} u_x &= \bar{e}_x + e_1 x + (e_3 - 2\bar{e}_\theta)\frac{y}{2} + e_4 \frac{2xy}{b}, \\ u_y &= \bar{e}_y + e_2 y + (e_3 + 2\bar{e}_\theta)\frac{x}{2} + e_5 \frac{2xy}{a} \end{aligned}\right\} \tag{4.10}$$

where $\bar{e}_x$, $\bar{e}_y$, $(e_3 - 2\bar{e}_\theta)$ and $(e_3 + 2\bar{e}_\theta)$ are constants of integration. From (4.3) it follows that

$$\gamma_{xy} = e_3 + e_4 \frac{2x}{b} + e_5 \frac{2y}{a}. \tag{4.11}$$

The three parameters $\bar{e}_x$, $\bar{e}_y$, $\bar{e}_\theta$ appear in (4.10) but not in the expressions for the strains (4.9) and (4.11). They represent the rigid-body translation and rotation of the centroid of the element.

The displacement function (4.10) has the right number of parameters (eight) to fit any arbitrary set of four nodal displacements. Equations (4.9) and (4.11) may be written in matrix form as

$$\epsilon(x,y) = \begin{bmatrix} 1 & 0 & 0 & 2y/b & 0 \\ 0 & 1 & 0 & 0 & 2x/a \\ 0 & 0 & 1 & 2x/b & 2y/a \end{bmatrix} \begin{bmatrix} e_1 \\ e_2 \\ e_3 \\ e_4 \\ e_5 \end{bmatrix} \tag{4.12a}$$

or as

$$\epsilon(x,y) = (\boldsymbol{B}(x,y))^t \boldsymbol{e}. \tag{4.12b}$$

From equation (4.1) we have

$$\boldsymbol{\sigma}(x,y) = \boldsymbol{D}\boldsymbol{\epsilon} = \boldsymbol{D}\boldsymbol{B}^t\boldsymbol{e} \tag{4.13}$$

where $\boldsymbol{D}$ is given by equation (4.2)

Although we have not defined $\boldsymbol{r}$ explicitly, the components of $\boldsymbol{r}$ are already defined by implication, due to the requirement that $\boldsymbol{r}$ must correspond to $\boldsymbol{e}$ in a work sense. If we consider an arbitrary virtual displacement defined by a change $\boldsymbol{e}^*$, then $\boldsymbol{e}^{*t}\boldsymbol{r}$ must equal the virtual work done by the stresses $\boldsymbol{\sigma}$ during the displacement. Thus

$$\boldsymbol{e}^{*t}\boldsymbol{r} = \int \boldsymbol{\epsilon}^{*t}\boldsymbol{\sigma}\, dV$$

the region of integration being the volume of the element. Now $\boldsymbol{\epsilon}^{*t} = \boldsymbol{e}^{*t}\boldsymbol{B}$ from (4.12b), so that substituting for $\boldsymbol{\sigma}$ from (4.13) we obtain

$$\boldsymbol{e}^{*t}\boldsymbol{r} = h \int\int \boldsymbol{e}^{*t}\boldsymbol{B}\boldsymbol{D}\boldsymbol{B}^t\boldsymbol{e}\, dx\, dy,$$

and since $\boldsymbol{e}$ is independent of x and y this may be written

$$\boldsymbol{e}^{*t}\boldsymbol{r} = \boldsymbol{e}^{*t}\left[h\iint \boldsymbol{BDB}^t\,dx\,dy\right]\boldsymbol{e}.$$

Finally, since $\boldsymbol{e}^*$ represents an arbitrary virtual displacement, we may "cancel" $\boldsymbol{e}^{*t}$ from this equation to give $\boldsymbol{r} = \boldsymbol{Ke}$, where

$$\boldsymbol{K} = h\iint \boldsymbol{BDB}^t dxdy. \tag{4.14}$$

Although derived for a simple rectangle, this formula holds for any element, provided that the deformation function $\boldsymbol{B}$ is defined appropriately. In this example the matrix product $\boldsymbol{BDB}^t$ is a 5×5 matrix whose elements are at most quadratic in x and y, and most of the integrals are trivially zero. Carrying out the integrations we obtain

$$\boldsymbol{K} = \frac{EV}{1-\nu^2}\begin{bmatrix} 1 & \nu & 0 & 0 & 0 \\ \nu & 1 & 0 & 0 & 0 \\ 0 & 0 & \dfrac{1-\nu}{2} & 0 & 0 \\ 0 & 0 & 0 & \dfrac{1}{3}\left(1+\dfrac{1-\nu}{2}\dfrac{a^2}{b^2}\right) & 0 \\ 0 & 0 & 0 & 0 & \dfrac{1}{3}\left(1+\dfrac{1-\nu}{2}\dfrac{b^2}{a^2}\right) \end{bmatrix} \tag{4.15}$$

The first three rows and columns of $\boldsymbol{K}$ are identical with the complete $\boldsymbol{K}$ matrix for the triangular element. This is what we should expect, since $\boldsymbol{e}_1$, $\boldsymbol{e}_2$ and $\boldsymbol{e}_3$ have exactly the same meaning as before. The inverse matrix $\boldsymbol{F}$ is essentially the matrix $\boldsymbol{F}$ given in (4.5b), with two additional diagonal elements.

The next step is to find the matrix $\boldsymbol{H}$. The direct approach via the equilibrium of the element requires a physical interpretation of the components of $\boldsymbol{r}$. Although in a sense this is desirable anyway as an aid to understanding the method, we can postpone the problem by a "compatibility" approach, in which we find $\boldsymbol{H}^t$ rather than $\boldsymbol{H}$.

If we return to (4.10) and substitute the nodal coordinates $x = \pm\, a/2$, $y = \pm\, b/2$, we obtain expressions for the components of the nodal displacements which may be written in matrix form as

$$\begin{bmatrix} \delta_{x1} \\ \delta_{y1} \\ \delta_{x2} \\ \delta_{y2} \\ \delta_{x3} \\ \delta_{y3} \\ \delta_{x4} \\ \delta_{y4} \end{bmatrix} = \begin{bmatrix} a/2 & 0 & b/4 & a/2 & 0 & 1 & 0 & -b/2 \\ 0 & b/2 & a/4 & 0 & b/2 & 0 & 1 & a/2 \\ -a/2 & 0 & b/4 & -a/2 & 0 & 1 & 0 & -b/2 \\ 0 & b/2 & -a/4 & 0 & -b/2 & 0 & 1 & -a/2 \\ -a/2 & 0 & -b/4 & a/2 & 0 & 1 & 0 & b/2 \\ 0 & -b/2 & -a/4 & 0 & b/2 & 0 & 1 & -a/2 \\ a/2 & 0 & -b/4 & -a/2 & 0 & 1 & 0 & b/2 \\ 0 & -b/2 & a/4 & 0 & -b/2 & 0 & 1 & a/2 \end{bmatrix} \begin{bmatrix} e_1 \\ e_2 \\ e_3 \\ e_4 \\ e_5 \\ \bar{e}_x \\ \bar{e}_y \\ \bar{e}_\theta \end{bmatrix}$$

Inverting this we obtain

$$\begin{bmatrix} e_1 \\ {}_2 \\ {}_3 \\ e_4 \\ e_5 \\ \hline \bar{e}_x \\ \bar{e}_y \\ \bar{e}_\theta \end{bmatrix} = \frac{1}{2ab} \left[\begin{array}{cc|cc|cc|cc} b & 0 & -b & 0 & -b & 0 & b & 0 \\ 0 & a & 0 & a & 0 & -a & 0 & -a \\ a & b & a & -b & -a & -b & -a & b \\ b & 0 & -b & 0 & b & 0 & -b & 0 \\ 0 & a & 0 & -a & 0 & a & 0 & -a \\ \hline ab/2 & 0 & ab/2 & 0 & ab/2 & 0 & ab/2 & 0 \\ 0 & ab/2 & 0 & ab/2 & 0 & ab/2 & 0 & ab/2 \\ -a/2 & b/2 & -a/2 & -b/2 & a/2 & -b/2 & a/2 & b/2 \end{array}\right] \begin{bmatrix} \delta_{x1} \\ \delta_{y1} \\ \delta_{x2} \\ \delta_{y2} \\ \delta_{x3} \\ \delta_{y3} \\ \delta_{x4} \\ \delta_{y4} \end{bmatrix} \tag{4.16}$$

The first five scalar equations of (4.16) give the required equation

$$\boldsymbol{e} = \boldsymbol{H}^t\boldsymbol{d} = \boldsymbol{H}_1{}^t\boldsymbol{d}_1 + \boldsymbol{H}_2{}^t\boldsymbol{d}_2 + \boldsymbol{H}_3{}^t\boldsymbol{d}_3 + \boldsymbol{H}_4{}^t\boldsymbol{d}_4.$$

The remaining three equations give the components of the rigid-body displacement of the centroid. As before, we regard the evaluation of the nodal stiffness matrices $\boldsymbol{H}_i\boldsymbol{K}\boldsymbol{H}_j{}^t$ as a task best done numerically by a computer.

The procedure we have described is a powerful algebraic method for finding $\boldsymbol{K}$ and $\boldsymbol{H}$ matrices. The reader may well feel, however, that it does not give much insight into the errors which may result from choosing a particular set of deformation patterns. We shall not attempt any detailed discussion of errors here. Instead we shall look a little more closely at one of the deformation modes we have chosen for our rectangle and try and get a physical picture of the stresses associated with it.

We consider the deformation associated with the component e_4. This is the mode in which ϵ_x varies linearly with y, as shown in Fig. 4.4a. From (4.9) and (4.11) we have $\epsilon_x = 2e_4y/b$, $\epsilon_y = 0$, $\gamma_{xy} = 2e_4x/b$. Corresponding to these strains, equation (4.1) gives us the corresponding stresses.

$$\sigma_x = \frac{E}{1-\nu^2} \cdot \frac{2e_4y}{b}, \quad \sigma_y = \frac{\nu E}{1-\nu^2} \cdot \frac{2e_4y}{b}, \quad \tau_{xy} = \frac{E}{2(1+\nu)} \cdot \frac{2e_4x}{b}.$$

Substituting the appropriate values for x and y gives the boundary stresses shown in Fig. 4.4b.

If we consider the quadrant $A1B$ it is clear that the nodal load component p_{x1} must be statically equivalent to the sum of the boundary stresses in the x-direction (the direct stress on $A1$ and the sheer stress on $B1$). If we consider a virtual deformation defined by a change $e_4{}^*$, the virtual work equation for the whole element is $p_{x1}\delta_{x1}{}^* + \ldots + p_{x4}\delta_{x4}{}^* = r_4e_4{}^* =$ virtual work done by the boundary stresses. From symmetry we have $\delta_{x1}{}^* = -\delta_{x2}{}^* = -\delta_{x3}{}^* = -\delta_{x4}{}^* = (a/2)e_4{}^*$ and $p_{x1} = -p_{x2} = p_{x3} = -p_{x4}$. It follows that $p_{x1} = r_4/2a$. Thus the component r_4 of $\boldsymbol{r}$ is associated with a self-equilibrating set of nodal forces $p_{x1} = -p_{x2} = p_{x3} = -p_{x4} = r_4/2a$. Evaluating the work done by the boundary stresses gives

$$r_4 = \frac{EV}{3(1-\nu^2)}\left[1 + \frac{1-\nu}{2} \cdot \frac{a^2}{b^2}\right] e_4$$

which agrees with equation (4.15).

Having identified r_4 with a specific set of nodal forces it is easy to write

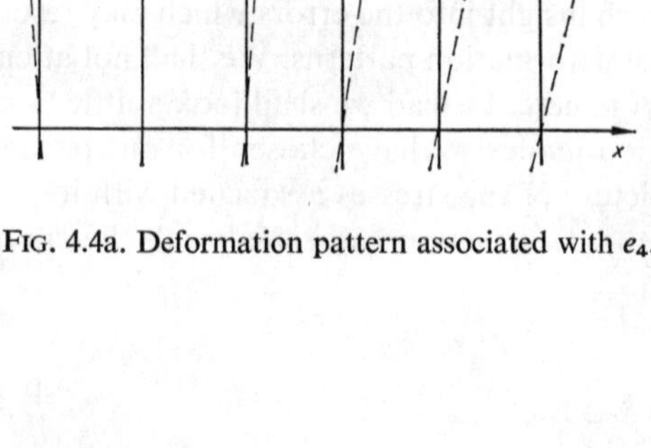

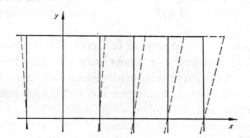

FIG. 4.4a. Deformation pattern associated with e_4.

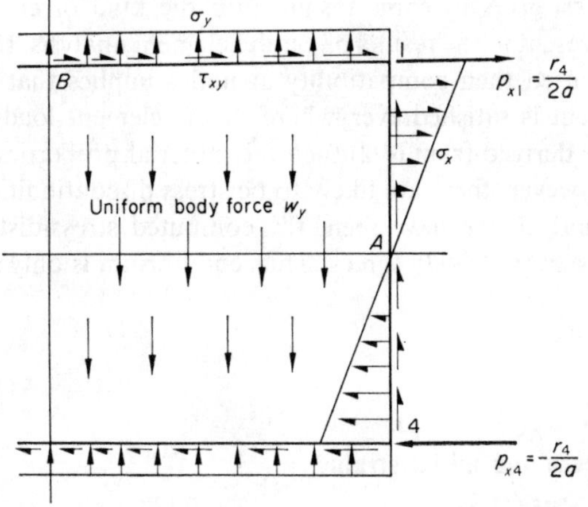

FIG. 4.4b. Boundary forces, body forces and equivalent nodal loads associated with e_4.

down the column of $\boldsymbol{H}$ associated with r_4. The reader will find that this column, derived by equilibrium considerations, is identical with the fourth row of $\boldsymbol{H}^t$ in (4.16).

Up to this point our analysis has merely confirmed the results already obtained by the formal algebraic method. However, substitution of the stresses given above into equation (4.4) reveals that the stress system associated with e_4 does *not* satisfy the conditions of equilibrium. In fact a deformation of the specified form requires a uniform body force

$$w_y = -\frac{\partial \sigma_y}{\partial y} - \frac{\partial \tau_{xy}}{\partial x} = -\frac{e_4 E}{b(1-\nu)} \quad \text{per unit volume}$$

to be applied to the material of the element, as shown in Fig. 4.4b. It is easy to verify that the resultant force due to the direct stress σ_y on 1 . . . 2 and 3 . . . 4, and the shear stress τ_{xy} on 2 . . . 3 and 1 . . . 4, exactly balances the integral of the body force w_y over the volume. The element is therefore in overall equilibrium, as we should expect.

This analysis gives us some insight into the kind of errors which are likely to be present in the results of a finite element analysis. If conformable elements are used then compatibility at nodes implies that compatibility of displacement is satisfied everywhere. If the element load/displacement equations are derived from (4.2) then the material properties are correctly modelled. However, there are likely to be stress discontinuities on element boundaries and, as we have seen, the computed stress-distribution may imply the existence of body forces. Thus equilibrium is only satisfied in an overall sense.

4.4. Body forces and initial strains

In Section 3.10 we developed a general technique for replacing distributed loads on a line element by equivalent loads at its ends. We now consider the same problem for area elements and describe a systematic technique for carrying out this process of replacement.

We take as our example a plane sheet of material divided into rectangular elements similar to the one discussed in the previous section. We imagine that a certain element carries a distributed load in the xy plane of $\boldsymbol{w}(x,y)$ per unit volume. Following the procedure of Section 3.10 we consider the loading on the system as the sum of

(a) The external applied load $\boldsymbol{w}(x,y)$, plus a set of loads $-\boldsymbol{p}_{\text{equiv}}$ at the nodes of the element, whose components are of such magnitudes as to prevent any displacement of the nodes.
(b) A set of loads $\boldsymbol{p}_{\text{equiv}}$ at the nodes of the element. This is the set used in the finite element analysis of the complete sheet.

The problem is to calculate $\boldsymbol{p}_{\text{equiv}}$ for the given distribution $\boldsymbol{w}(x,y)$. We note that $\boldsymbol{w}(x,y)$ and $-\boldsymbol{p}_{\text{equiv}}$ form a system of forces in equilibrium. This suggests an application of the principle of virtual work on the same lines as that used in the previous section.

We first write the displacement expressions (4.10) in matrix form as

$$\begin{bmatrix} u_x \\ u_y \end{bmatrix} = \begin{bmatrix} x & 0 & y/2 & 2xy/b & 0 & 1 & 0 & -y \\ 0 & y & x/2 & 0 & 2xy/a & 0 & 1 & x \end{bmatrix} \begin{bmatrix} e_1 \\ e_2 \\ e_3 \\ e_4 \\ e_5 \\ \bar{e}_x \\ \bar{e}_y \\ \bar{e}_\theta \end{bmatrix} \tag{4.17a}$$

or

$$\boldsymbol{u}(x,y) = (\boldsymbol{A}(x,y))^t \boldsymbol{e}_+ \tag{4.17b}$$

where $\boldsymbol{e}_+$ denotes the extended vector formed by adding the rigid-body displacement components $\bar{e}_x$, $\bar{e}_y$ and $\bar{e}_\theta$ to the deformation vector $\boldsymbol{e}$. We now impose a virtual displacement on the system defined by a change $\boldsymbol{e}_+^*$. This produces displacements $\boldsymbol{u}^*$ in the element material and displacements $\boldsymbol{d}^*$ at the nodes. The virtual work done by all the forces acting on the element is therefore

$$\int \boldsymbol{u}^{*t} w dV - \boldsymbol{d}^{*t} \boldsymbol{p}_{\text{equiv}} = 0.$$

On substitution from (4.17b) this becomes

$$\boldsymbol{e}_+^{*t} h \left[\iint \boldsymbol{A} w dx dy \right] - \boldsymbol{d}^{*t} \boldsymbol{p}_{\text{equiv}} = 0. \tag{4.18}$$

If we now write (4.16) as

$$\boldsymbol{e}_+ = \boldsymbol{H}_+{}^t \boldsymbol{d} \tag{4.19}$$

and use this equation to substitute for $\boldsymbol{e}_+^{*t}$ in equation (4.18) we obtain

$$\boldsymbol{d}^{*t} \boldsymbol{H}_+ h \left[\iint \boldsymbol{A} w dx dy \right] - \boldsymbol{d}^{*t} \boldsymbol{p}_{\text{equiv}} = 0$$

or since $\boldsymbol{d}^*$ is arbitrary

$$\boldsymbol{p}_{\text{equiv}} = \boldsymbol{H}_+ h \iint \boldsymbol{A} w dx dy. \tag{4.20}$$

Note that this analysis determines the nodal loading vector which is equivalent to the distributed loading $w(x,y)$ as far as the rest of the system is concerned. There is no guarantee that the deformation patterns we have chosen for the element are appropriate for finding the stresses actually present *within* the element under the given loading $\boldsymbol{w}$.

As a simple example of this analysis consider a uniform loading $\boldsymbol{w}(x,y) = \begin{bmatrix} w \\ 0 \end{bmatrix}$. The integrand $\boldsymbol{A}\boldsymbol{w}$ in (4.20) is equal to

$$\begin{bmatrix} x & 0 \\ 0 & y \\ y/2 & x/2 \\ 2xy/b & 0 \\ 0 & 2xy/a \\ 1 & 0 \\ 0 & 1 \\ -y & x \end{bmatrix} \begin{bmatrix} w \\ 0 \end{bmatrix} = \begin{bmatrix} wx \\ 0 \\ wy/2 \\ 2wxy/b \\ 0 \\ w \\ 0 \\ -wy \end{bmatrix}, \text{ which integrates to } \begin{bmatrix} 0 \\ 0 \\ 0 \\ 0 \\ 0 \\ wab \\ 0 \\ 0 \end{bmatrix}$$

Multiplying this by $\boldsymbol{H}_+h$ (only the sixth column of $\boldsymbol{H}_+$, i.e. the sixth row of $\boldsymbol{H}_+{}^t$, is relevant) gives

$$\boldsymbol{p}_{\text{equiv}} = \begin{bmatrix} wab/4 \\ 0 \\ wab/4 \\ 0 \\ wab/4 \\ 0 \\ wab/4 \\ 0 \end{bmatrix}$$

which is precisely what we should expect intuitively.

Initial strains, such as those due to temperature, also give rise to equivalent nodal forces. The analysis follows very much the same pattern as that already given for body forces, the main difference being that the nodal forces associated with initial strains are essentially self-equilibrating.

Once again we consider a plane sheet divided up into elements, one of which has some strains $\boldsymbol{\epsilon}_0(x,y)$ when unstressed. We have to find $\boldsymbol{p}_{\text{equiv}}$, where $-\boldsymbol{p}_{\text{equiv}}$ is the vector of nodal loads which will prevent any nodal displacements of the element.

The stress distribution in the element is now given by

$$\boldsymbol{\sigma} = \boldsymbol{D}(\boldsymbol{\epsilon} - \boldsymbol{\epsilon}_0).$$

The actual strain $\boldsymbol{\epsilon}$ can be expressed by (4.12b) as a function of the deformation vector $\boldsymbol{e}$,

$$\boldsymbol{\epsilon} = \boldsymbol{B}^t\boldsymbol{e}.$$

If the nodal displacements are all zero then $\boldsymbol{e}$ is zero and the actual strains $\boldsymbol{\epsilon}$ are everywhere zero. Thus we have a system of equivalent stresses $\boldsymbol{\sigma} = -\boldsymbol{D}\boldsymbol{\epsilon}_0$ which must be in equilibrium with the nodal loads $-\boldsymbol{p}_{\text{equiv}}$. Once again we apply the principle of virtual work, considering a virtual deformation defined by a change $\boldsymbol{e}^*$. Equating the work done by $-\boldsymbol{p}_{\text{equiv}}$ to the work done by $\boldsymbol{\sigma}$ gives

$$-\boldsymbol{d}^{*t}\boldsymbol{p}_{\text{equiv}} = \int \boldsymbol{\epsilon}^{*t}\boldsymbol{\sigma}\,dV$$

which gives

$$\boldsymbol{d}^{*t}\boldsymbol{p}_{\text{equiv}} = h\iint \boldsymbol{e}^{*t}\boldsymbol{B}\boldsymbol{D}\boldsymbol{\epsilon}_0\,dxdy.$$

In the same way as before we use the fact that $\boldsymbol{e}^{*t} = \boldsymbol{d}^{*t}\boldsymbol{H}$ to obtain finally

$$\boldsymbol{p}_{\text{equiv}} = \boldsymbol{H}h\iint \boldsymbol{B}\boldsymbol{D}\boldsymbol{\epsilon}_0\,dxdy. \tag{4.21}$$

If we compare this analysis with the analysis given earlier for body-forces we see that in the body-force analysis we used a virtual displacement $\boldsymbol{e}_+^*$ which involved rigid-body displacement components as well as deformations. This is due to the fact that body-forces are not, in general, self-equilibrating. Equations (4.20) and (4.21) hold for any plane element, provided the various vectors and matrices are defined appropriately.

As an example of this analysis, consider a rectangular element which experiences a uniform temperature rise θ. The strain vector $\boldsymbol{\epsilon}_0(x,y)$ is constant and equals

$$\begin{bmatrix}\alpha\theta\\ \alpha\theta\\ 0\end{bmatrix}, \text{ so that the product } \boldsymbol{D}\boldsymbol{\epsilon}_0 \text{ is equal to } \begin{bmatrix}\sigma_0\\ \sigma_0\\ 0\end{bmatrix},$$

where $\sigma_0 = E\alpha\theta/(1 - \nu)$. Using the expression for $\boldsymbol{B}$ given in (4.12) we obtain

$$BD\epsilon_0 = \sigma_0 \begin{bmatrix} 1 \\ 1 \\ 0 \\ 2y/b \\ 2x/a \end{bmatrix}$$

Integrating and multiplying by Hh gives finally

$$p_{\text{equiv}} = \frac{h\sigma_0}{2} \begin{bmatrix} b \\ a \\ -b \\ a \\ -b \\ -a \\ b \\ a \end{bmatrix}$$

This set of nodal loads is illustrated in Fig. 4.5. Again it agrees with what one would intuitively expect (and indeed could derive by a much less general argument).

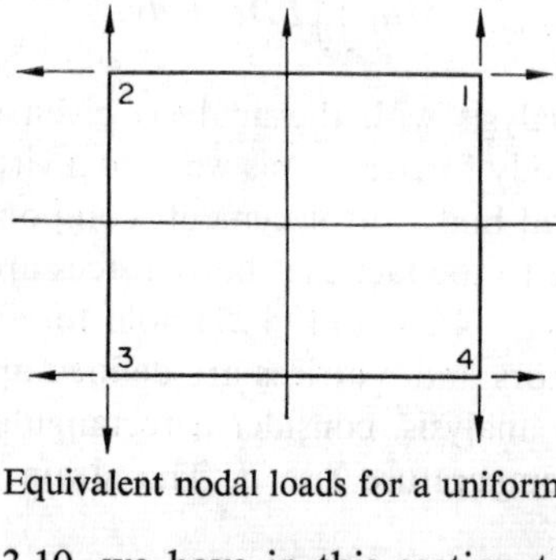

FIG. 4.5. Equivalent nodal loads for a uniform temperature rise.

As in Section 3.10, we have in this section treated the calculation of equivalent nodal forces as something separate from the actual process of analysis. An alternative approach is to include the vector p_{equiv} in the nodal load/displacement equations, writing them as

$$p_i = H_i K H_j^t d_j - (p_{\text{equiv}})_i. \tag{4.22}$$

If this is done then the equivalent load vectors for the elements are automatically included in the load vector for the complete system when the load/displacement equations are assembled. (See Section 5.3 for a further discussion of this point.)

4.5. Transformation of axes

The procedure for transforming the load/displacement equations into global coordinates follows the pattern set out in Section 2.4. In plane stress problems each node has two degrees of freedom, so that the transformation equations are

$$\boldsymbol{p}_i' = \boldsymbol{T}\boldsymbol{p}_i, \quad \boldsymbol{d}_i = \boldsymbol{T}^t\boldsymbol{d}_i'$$

where

$$\boldsymbol{T} = \begin{bmatrix} \cos\alpha & -\sin\alpha \\ \sin\alpha & \cos\alpha \end{bmatrix}.$$

The element load/displacement equations $\boldsymbol{p}_i = \boldsymbol{H}_i\boldsymbol{K}\boldsymbol{H}_j\boldsymbol{d}_j$ transform into

$$\boldsymbol{p}_i' = \boldsymbol{T}\boldsymbol{H}_i\boldsymbol{K}\boldsymbol{H}_j^t\boldsymbol{T}^t\boldsymbol{d}_j'. \tag{4.23}$$

This may be written in the usual way as $\boldsymbol{p}_i' = \boldsymbol{T}\boldsymbol{K}_{ij}\boldsymbol{T}^t\boldsymbol{d}_j' = \boldsymbol{K}_{ij}'\boldsymbol{d}_j'$. Alternatively we may put $\boldsymbol{H}_i' = \boldsymbol{T}\boldsymbol{H}_i$, so that the equations become

$$\boldsymbol{p}_i' = \boldsymbol{H}_i'\boldsymbol{K}\boldsymbol{H}_j'^t\boldsymbol{d}_j'. \tag{4.24}$$

Note that in these transformations we do not attempt to transform the vectors $\boldsymbol{r}$, $\boldsymbol{e}$, $\boldsymbol{\sigma}$ or $\boldsymbol{\epsilon}$. As mentioned in Section 2.2, these quantities may only be treated as vectors in a particular set of axes. They do not, in general, transform like "geometrical vectors" when coordinate axes are changed.

4.6. The replacement of distributed mass

The treatment of distributed mass in continuum vibration problems follows much the same pattern as that described in Section 3.11. As in that section, we assume that the displacement of an arbitrary point in a vibrating element is the same as it would be in the static case with no body forces and the same nodal displacements. For a continuum element this deformation is already defined by an equation such as (4.17b). If the density of the element material is constant and equal to ρ the inertia loading $\boldsymbol{w}(x,y)$ is given by

$$\boldsymbol{w}(x,y) = -\rho\ddot{\boldsymbol{u}}(x,y) = -\rho\boldsymbol{A}^t\ddot{\boldsymbol{e}}_+.$$

Combining this with (4.19) we obtain $\boldsymbol{w}(x,y) = -\rho\boldsymbol{A}^t\boldsymbol{H}_+{}^t\ddot{\boldsymbol{d}}$. Substituting this in (4.20) gives

$$\boldsymbol{p}_{\text{equiv}} = -\rho h\boldsymbol{H}_+ \iint \boldsymbol{A}\boldsymbol{A}^t dxdy\boldsymbol{H}_+{}^t\ddot{\boldsymbol{d}}. \qquad (4.25a)$$

Following Section 3.11 we write this as

$$(\boldsymbol{p}_{\text{equiv}})_i = -\boldsymbol{M}_{ij}\ddot{\boldsymbol{d}}_j \qquad (4.25b)$$

where

$$\boldsymbol{M}_{ij} = \rho h(\boldsymbol{H}_+)_i \iint \boldsymbol{A}\boldsymbol{A}^t dxdy(\boldsymbol{H}_+{}^t)_j. \qquad (4.25c)$$

The mass matrices $\boldsymbol{M}_{ij}$ transform into global coordinates in exactly the same way as the element stiffness matrices $\boldsymbol{K}_{ij}$.

Equations (4.25) hold for any plane element, provided that the matrices $\boldsymbol{H}_+$ and $\boldsymbol{A}$ are defined appropriately. For the rectangular element of Fig. 3.3 the matrices $\boldsymbol{M}_{ij}$ are

$$\frac{M}{36}\begin{bmatrix}4 & 0\\ 0 & 4\end{bmatrix}, (j=i): \frac{M}{36}\begin{bmatrix}2 & 0\\ 0 & 2\end{bmatrix}, (j=i\pm 1,3): \frac{M}{36}\begin{bmatrix}1 & 0\\ 0 & 1\end{bmatrix}, (j=i\pm 2).$$

where M is the mass of the element. Details of mass matrices for other types of element will be found in Przemieniecki (1968) and Zienkiewicz (1971).

4.7. More complex elements

One may argue that the rectangular element is better for plane stress analysis than the triangle because it allows more variation in stress and strain within the element. This improvement is due to the increase in the number of independent deformation parameters in the vector $\boldsymbol{e}$ from 3 to 5. It is clear that with more deformation parameters the computed stress distribution is likely to be an even closer approximation to the true stress distribution.

One may achieve an increase in the number of deformation parameters by increasing the number of nodes. For example, consider the triangle shown in Fig. 4.6. This has six nodes, twelve nodal degrees of freedom and

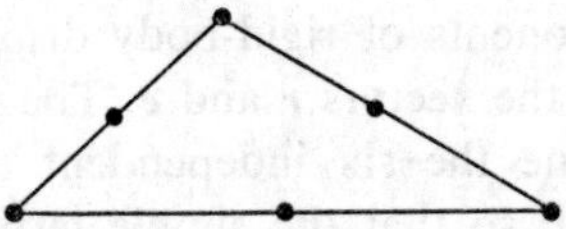

FIG. 4.6. A plane triangle with six nodes.

therefore nine independent deformation parameters in the vector ***e***. The nodes at the mid-points of the sides are useful in ensuring that the element is still conformable. With only two nodes on a straight boundary we argued that displacements must vary linearly along each boundary for the element to be conformable. With three nodes we can allow parabolic variation, since two parabolae passing through the same three points must be identical. A similar argument applies to the boundaries of quadrilateral elements.

In discussing the rectangular element we stated that any quadrilateral could be mapped on to a rectangle. This is also the case for elements with curved boundaries—Fig. 4.7 shows a typical curvilinear element. Elements

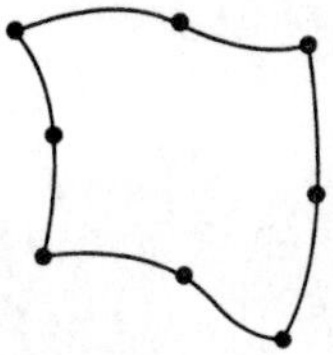

FIG. 4.7. An isoparametric element.

of this sort are termed *isoparametric* elements and were first introduced by Irons in 1966. Details will be found in Zienkiewicz (1971). The general procedure of Section 4.3 for determining the element stiffness matrix ***K*** can still be followed in the case of more complex elements, but the integrals which occur usually have to be evaluated numerically.

So far we have been concerned with problems of plane stress. The properties of three-dimensional elements may be investigated by very similar techniques. For example, the simplest three-dimensional element is the tetrahedron. This has four vertices, three degrees of freedom at each vertex and therefore twelve nodal degrees of freedom in all. There are six equations of equilibrium to be satisfied by the nodal forces (and, equivalently, six components of rigid-body displacement), so that there are six components in the vectors $\boldsymbol{r}$ and $\boldsymbol{e}$. The six components of $\boldsymbol{e}$ are only sufficient to define the six independent components of a three-dimensional stress field, so that the simple tetrahedron, like the simple triangle, is a constant-stress element. The next elements in order of complexity are the triangular prism (six nodes) and the rectangular brick (eight nodes). Both of these may be analysed by techniques very similar to those used in Section 4.3. As with plane elements, more complicated elements with extra boundary nodes and curvilinear edges have been developed.

An important area of finite element analysis is concerned with the development of elements for plates and shells. These elements are in a sense similar to plane elements, but have displacements normal to the plane of the element. Details of these elements may be found in the standard texts.

CHAPTER 5

The Equilibrium or Displacement Method

In the previous two chapters we considered the properties of individual structural elements. We described techniques for computing the ***H*** and ***K*** matrices for various types of element, and from these matrices we derived the corresponding nodal stiffness matrices $\boldsymbol{K}_{ij}$. We also showed how the nodal load/displacement equations can be transformed into an arbitrary set of global coordinates. Finally we showed how distributed mass, loading or initial strains can be replaced by equivalent loads concentrated at the nodes.

In this chapter we describe a systematic procedure for assembling the load/displacement equations of a complete structure from the load/displacement equations of its individual elements. It is this procedure which constitutes what is usually termed the "equilibrium" or "displacement" method. As with other techniques based on a simple idea, it is difficult to know who should be credited with the invention of the method. The idea is certainly present in a paper by Bendixen and Ostenfeld (Ostenfeld, 1926). In a sense, however, the process only becomes identifiable as a separate "method" when written out in a systematic general notation. One of the earliest writers to develop such a notation was Kron (1944) and the presentation given here owes a great deal to his work.

In deriving the coefficients of the matrices $\boldsymbol{K}_{ij}$ we were concerned with the precise form of a structural element—whether it was a line, area or volume element, its elastic material properties and its orientation to the global coordinate axes. We also had to introduce various assumptions and approximations to make the equations linear. However, we did not have to worry about the complexity of the structure of which the element might eventually be a part. Now, when we come to assemble the equations for a complete structure, we find that we do not need to consider the physical nature of the elements, but only the way in which they are connected together. The procedure we develop in this chapter is applicable whether our "structure" is an assembly of triangular plane elements or a rigid-jointed space frame.

Although our aim is to develop general rules of procedure it is easiest to introduce the method by considering examples in which the elements are as simple as possible. In this chapter our examples are all skeletal structures—mostly pin-jointed or rigid-jointed plane frameworks. The reader interested in applications of the method to continuum analysis should not find it difficult to develop similar examples using the triangular and rectangular elements described in Chapter 4.

5.1. The analysis of a plane pin-jointed truss

For our first example we consider the plane pin-jointed truss shown in Fig. 5.1. The arrow placed on each bar indicates the direction of the positive x-axis for the corresponding member coordinate system, and hence identifies the ends 1 and 2 of the bar. We assume that the joint loads $\boldsymbol{p}_A, \boldsymbol{p}_B$ are known.

For each bar in the truss the nodal load/displacement equations are given in the global coordinate system $x'y'$ by equation (3.20a),

$$\left.\begin{aligned} \boldsymbol{p}_1' &= \boldsymbol{K}'\boldsymbol{d}_1' - \boldsymbol{K}'\boldsymbol{d}_2', \\ \boldsymbol{p}_2' &= -\boldsymbol{K}'\boldsymbol{d}_1' + \boldsymbol{K}'\boldsymbol{d}_2', \end{aligned}\right\} \tag{5.1}$$

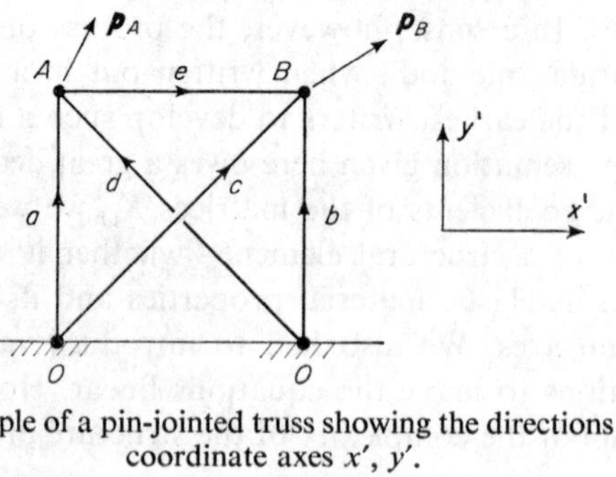

FIG. 5.1. An example of a pin-jointed truss showing the directions of the global coordinate axes x', y'.

where $\boldsymbol{K}'$ is the 2×2 matrix defined in (3.20b) and each vector $\boldsymbol{p}'$ and $\boldsymbol{d}'$ has two components. Applying the condition that the displacements of the ends of members must be equal to the displacements of the appropriate joints we have the following equations:

$$\left.\begin{array}{lll} \text{Member } a & \boldsymbol{d}_1' = \boldsymbol{0} & \boldsymbol{d}_2' = \boldsymbol{d}_A, \\ \text{Member } b & \boldsymbol{d}_1' = \boldsymbol{0} & \boldsymbol{d}_2' = \boldsymbol{d}_B, \\ \text{Member } c & \boldsymbol{d}_1' = \boldsymbol{0} & \boldsymbol{d}_2' = \boldsymbol{d}_B, \\ \text{Member } d & \boldsymbol{d}_1' = \boldsymbol{0} & \boldsymbol{d}_2' = \boldsymbol{d}_A, \\ \text{Member } e & \boldsymbol{d}_1' = \boldsymbol{d}_A & \boldsymbol{d}_2' = \boldsymbol{d}_B. \end{array}\right\} \tag{5.2}$$

Substituting these expressions for the member end-displacements into equations (5.1) for each member gives

$$\left.\begin{array}{lll} \text{Member } a & \boldsymbol{p}_{2a}' & = \boldsymbol{K}_a'\boldsymbol{d}_A, \\ \text{Member } b & \boldsymbol{p}_{2b}' & = \boldsymbol{K}_b'\boldsymbol{d}_B, \\ \text{Member } c & \boldsymbol{p}_{2c}' & = \boldsymbol{K}_c'\boldsymbol{d}_B, \\ \text{Member } d & \boldsymbol{p}_{2d}' & = \boldsymbol{K}_d'\boldsymbol{d}_A, \\ \text{Member } e & \boldsymbol{p}_{1e}' & = \boldsymbol{K}_e'\boldsymbol{d}_A - \boldsymbol{K}_e'\boldsymbol{d}_B, \\ & \boldsymbol{p}_{2e}' & = -\boldsymbol{K}_e'\boldsymbol{d}_A + \boldsymbol{K}_e'\boldsymbol{d}_B. \end{array}\right\} \tag{5.3}$$

The expressions for $\boldsymbol{p}_1'$ have been omitted in the case of members *a*, *b*, *c* and *d*, since they are merely foundation reactions and do not enter into the joint equilibrium equations. The primes are omitted from $\boldsymbol{d}_A$ and $\boldsymbol{d}_B$, since these can only be defined in global coordinates.

The conditions of joint equilibrium in this example are simply

$$\left.\begin{array}{l} \boldsymbol{p}_A = \boldsymbol{p}_{2a}' + \boldsymbol{p}_{2d}' + \boldsymbol{p}_{1e}', \\ \boldsymbol{p}_B = \boldsymbol{p}_{2b}' + \boldsymbol{p}_{2c}' + \boldsymbol{p}_{2e}', \end{array}\right\} \tag{5.4}$$

where the primes are omitted from $\boldsymbol{p}_A$ and $\boldsymbol{p}_B$ for the reason given above. Substituting into these equations from (5.3) gives

$$\begin{bmatrix} \boldsymbol{p}_A \\ \boldsymbol{p}_B \end{bmatrix} = \begin{bmatrix} \boldsymbol{K}_a' + \boldsymbol{K}_d' + \boldsymbol{K}_e' & -\boldsymbol{K}_e' \\ -\boldsymbol{K}_e' & \boldsymbol{K}_b' + \boldsymbol{K}_c' + \boldsymbol{K}_e' \end{bmatrix} \begin{bmatrix} \boldsymbol{d}_A \\ \boldsymbol{d}_B \end{bmatrix}. \tag{5.5}$$

The matrix in equation (5.5) is an example of a partitioned matrix, the coefficients in this case being 2×2 matrices. It is referred to as the *stiffness matrix* of the structure and is represented by the symbol $\mathbf{K}$.

Although the load/displacement equations for the individual members are singular, we argue on physical grounds that equation (5.5) must have a unique solution for a given loading, and that therefore the complete stiffness matrix of the structure must be non-singular. Solving the set of four simultaneous equations (5.5) we obtain the displacements $\boldsymbol{d}_A$, $\boldsymbol{d}_B$.

The final step is to compute the bar forces. The best procedure is to take each bar in turn and work out the end-displacements in the associated *member* coordinate system using the relationship (3.19b)

$$d_{1,2} = [\cos\alpha \quad \sin\alpha]\, \boldsymbol{d}_{1,2}'.$$

The axial force then follows from (3.18) as

$$p_2 = K(d_2 - d_1).$$

This is rather more efficient than computing the end-forces in global coordinates from (5.3) and then changing to member coordinates.

If a structure has to be analysed for a large number of different loadings it may be better to invert the stiffness matrix $\mathbf{K}$ to give the flexibility matrix $\mathbf{F}$, and then find the displacements for each loading from the equation $\mathbf{d} = \mathbf{Fp}$. With this procedure the matrix inversion only has to be carried out once; the subsequent analysis for any particular loading merely requires multiplication of the appropriate load vector $\mathbf{p}$ by the matrix $\mathbf{F}$. However, there is a practical drawback to this procedure, which is discussed in Section 10.1.

It is interesting to note the way in which the stiffness matrix in (5.5) reflects the arrangement of bars in the structure. The bars *a* and *d* merely contribute to the direct stiffness of joint *A* and consequently their $\boldsymbol{K}'$ matrices only appear in the *first* leading diagonal coefficient, which relates the displacement of joint *A* to the load applied there. Similarly the bars *b* and *c* merely contribute to the stiffness of joint *B*, and their $\boldsymbol{K}'$ matrices consequently appear only in the *second* leading diagonal coefficient. The bar *e*, on the other hand, connects joints *A* and *B*, and contributes to both the leading diagonal terms. It also provides the off-diagonal matrices $-\boldsymbol{K}_e'$ which constitute the mathematical link between the displacements $\boldsymbol{d}_A$ and $\boldsymbol{d}_B$, thus reflecting the physical link between the two joints which the bar provides in the structure.

This close relationship between the form of the stiffness matrix and the geometrical arrangement of the structure makes it possible to write down the complete matrix in terms of the member stiffness matrices merely by inspection, without formally writing out the conditions of equilibrium and compatibility. The general rule is that if a member s connects joints I and J then $\boldsymbol{K}_s'$ is added to the Ith and Jth leading diagonal sub-matrices, and $-\boldsymbol{K}_s'$ is placed in the corresponding off-diagonal positions (i.e. row I, column J, and row J, column I). If we wish to modify a design and remove a member from a frame, we merely remove its $\boldsymbol{K}'$ matrices from the appropriate places in the complete stiffness matrix, without having to alter the remaining terms. (However, this will in general have the effect of altering all the coefficients of the flexibility matrix $\mathbf{F}$.)

One of the most important features of the stiffness matrix of a structure is the fact that it is formed by adding up the stiffness matrices of the structure's component parts. If in Fig. 5.1 we prevent joint A from displacing we have the structure shown in Fig. 5.2, and equation (5.5) reduces to $\boldsymbol{p}_B = (\boldsymbol{K}_b' + \boldsymbol{K}_c' + \boldsymbol{K}_e')\boldsymbol{d}_B$. This equation may be regarded as a generalization of the scalar equation $p = (k_1 + k_2 + k_3)d$ for the force/displacement relationship of three springs of stiffness k_1, k_2, k_3 arranged in parallel, as shown in Fig. 5.3.

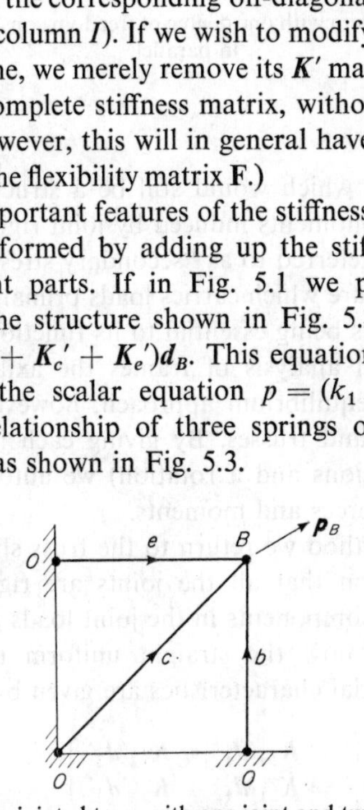

FIG. 5.2. A simple pin-jointed truss with one joint and two degrees of freedom.

5.2. The analysis of a plane rigid-jointed truss

In the analysis of plane structures by traditional methods it is usual to make a sharp distinction between trusses and frames, and to use different methods of analysis in the two cases. A truss is essentially a

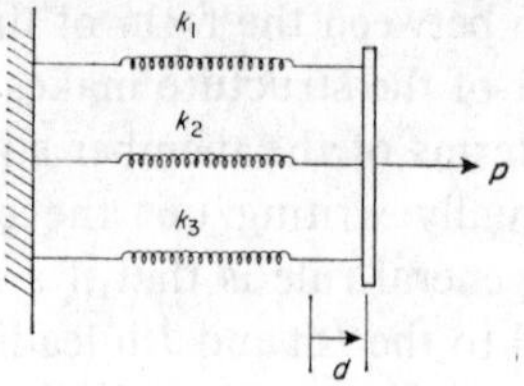

FIG. 5.3. A simple system with one degree of freedom consisting of three springs in parallel.

triangulated structure which would still be a structure if its joints were pinned. The bending moments induced by joint rigidity are usually small and indeed are often referred to as "secondary stresses". A frame, on the other hand, is a structure which carries loads primarily by bending action, the rigidity of its joints being essential to its function as a structure.

In the conventional analysis of frames the axial strains are usually ignored. The matrix equilibrium approach, however, makes no distinction between frames and trusses. By giving each joint three degrees of freedom (two translations and a rotation) we automatically include the effects of both axial forces and moments.

To illustrate the method we return to the truss shown in Fig. 5.1, now making the assumption that all the joints are rigid. (This implies the inclusion of moment components in the joint loads $\boldsymbol{p}_A$ and $\boldsymbol{p}_B$.) Our basic structural element is now the straight uniform member described in Section 3.2, whose nodal characteristics are given by equations (3.13)

$$\left.\begin{aligned} \boldsymbol{p}_1' &= \boldsymbol{K}_{11}'\boldsymbol{d}_1' + \boldsymbol{K}_{12}'\boldsymbol{d}_2', \\ \boldsymbol{p}_2' &= \boldsymbol{K}_{21}'\boldsymbol{d}_1' + \boldsymbol{K}_{22}'\boldsymbol{d}_2'. \end{aligned}\right\} \qquad (5.6)$$

The analysis follows very much the same pattern as has already been described in the pin-jointed case. The equations of joint compatibility (5.2) are unchanged, although the displacement vectors in those equations now have rotations included in their components. Substituting these expressions for the end-displacements of the members into equations (5.6) for the individual members gives

$$\left.\begin{aligned}
&\text{Member } a \quad \boldsymbol{p}_{2a}' = (\boldsymbol{K}_{22}')_a \boldsymbol{d}_A,\\
&\text{Member } b \quad \boldsymbol{p}_{2b}' = (\boldsymbol{K}_{22}')_b \boldsymbol{d}_B,\\
&\text{Member } c \quad \boldsymbol{p}_{2c}' = (\boldsymbol{K}_{22}')_c \boldsymbol{d}_B,\\
&\text{Member } d \quad \boldsymbol{p}_{2d}' = (\boldsymbol{K}_{22}')_d \boldsymbol{d}_A,\\
&\text{Member } e \quad \boldsymbol{p}_{1e}' = (\boldsymbol{K}_{11}')_e \boldsymbol{d}_A + (\boldsymbol{K}_{12}')_e \boldsymbol{d}_B,\\
&\qquad\qquad\quad \boldsymbol{p}_{2e}' = (\boldsymbol{K}_{21}')_e \boldsymbol{d}_A + (\boldsymbol{K}_{22}')_e \boldsymbol{d}_B.
\end{aligned}\right\} \tag{5.7}$$

where, as before, primes are omitted from the displacement vectors since these must obviously be defined in the global coordinate system. These equations are identical with (5.3) except for the suffixes which now have to be attached to the $\boldsymbol{K}_{ij}'$ matrices, and the fact that the end-loads now have moment components. As before, the expressions for $\boldsymbol{p}_1'$ have been omitted in the case of members a, b, c and d.

The conditions of joint equilibrium are also exactly the same as before,

$$\boldsymbol{p}_A = \boldsymbol{p}_{2a}' + \boldsymbol{p}_{2d}' + \boldsymbol{p}_{1e}',$$
$$\boldsymbol{p}_B = \boldsymbol{p}_{2b}' + \boldsymbol{p}_{2c}' + \boldsymbol{p}_{2e}'.$$

Substituting equations (5.7) into these equations gives

$$\begin{bmatrix}\boldsymbol{p}_A\\ \boldsymbol{p}_B\end{bmatrix} = \begin{bmatrix}(\boldsymbol{K}_{22}')_a + (\boldsymbol{K}_{22}')_d + (\boldsymbol{K}_{11}')_e & (\boldsymbol{K}_{12}')_e\\ (\boldsymbol{K}_{21}')_e & (\boldsymbol{K}_{22}')_b + (\boldsymbol{K}_{22}')_c + (\boldsymbol{K}_{22}')_e\end{bmatrix}\begin{bmatrix}\boldsymbol{d}_A\\ \boldsymbol{d}_B\end{bmatrix} \tag{5.8}$$

It will be seen that these equations are very similar to the equations (5.5) derived for the pin-jointed truss. As before, the form of the complete stiffness matrix reflects the arrangement of bars in the structure. The addition of the two joint rotations as extra degrees of freedom means that equation (5.8) represents six linear simultaneous equations for the six components of joint displacement. Once these equations have been solved, the internal forces and moments in the truss may be found from (5.7).

In this particular problem the equilibrium method has given us a complete secondary-stress analysis of a rigid-jointed truss. If we now remove the members c and d from the structure we obtain an unbraced portal frame, which would normally be analysed in quite a different manner. However, all we need to do to obtain a complete solution of the new problem (including axial strain effects) is to drop the matrices $(\boldsymbol{K}_{22}')_c$

and $(K_{22}')_d$ from equations (5.8), solve the resulting equations, and use the relevant equations from the set (5.7) to find the forces and moments acting on the ends of the remaining members a, b and e.

5.3. The assembly of the stiffness matrix of a structure

It should by now be clear that the assembly of the complete stiffness matrix of a structure from the individual member stiffness matrices depends only on the way the members are connected together. Once the load/displacement equations have been found for all the members and have been transformed into global coordinates, the application of the conditions of equilibrium and compatibility involves no further reference to the detailed characteristics of the elements, and may be carried out formally in terms of the $\boldsymbol{p}'$ and $\boldsymbol{d}'$ vectors and the $\boldsymbol{K}_{ij}'$ matrices of the members. It is true, of course, that these vectors will have different numbers of components in different cases, and that the coefficients of the $\boldsymbol{K}_{ij}'$ matrices will depend on the type of element involved, but as long as the equations are written in terms of vectors and matrices this does not affect their general form.

We now consider the problem of assembling the complete matrix $\mathbf{K}$ for a general structure formed from line elements. We imagine that the joints of the structure are denoted by $A, B, \ldots, N$, points of attachment to foundations being regarded as joint O. We shall assume that for each member end 1 corresponds to the "lower" and end 2 to the "higher" lettered joint. Consider now the contribution which a member r running from joint I to joint J $(I < J)$ makes to the final set of load/displacement equations. The equations for the members are

$$\boldsymbol{p}_{1r}' = (\boldsymbol{K}_{11}')_r \boldsymbol{d}_{1r}' + (\boldsymbol{K}_{12}')_r \boldsymbol{d}_{2r}',$$

$$\boldsymbol{p}_{2r}' = (\boldsymbol{K}_{21}')_r \boldsymbol{d}_{1r}' + (\boldsymbol{K}_{22}')_r \boldsymbol{d}_{2r}',$$

or, substituting the compatibility conditions $\boldsymbol{d}_{1r}' = \boldsymbol{d}_I$, $\boldsymbol{d}_{2r}' = \boldsymbol{d}_J$,

$$\left.\begin{aligned} \boldsymbol{p}_{1r}' &= (\boldsymbol{K}_{11}')_r \boldsymbol{d}_I + (\boldsymbol{K}_{12}')_r \boldsymbol{d}_J, \\ \boldsymbol{p}_{2r}' &= (\boldsymbol{K}_{21}')_r \boldsymbol{d}_I + (\boldsymbol{K}_{22}')_r \boldsymbol{d}_J. \end{aligned}\right\} \qquad (5.9a)$$

The equilibrium equations for joints I and J are

$$\left.\begin{aligned} \boldsymbol{p}_I &= \boldsymbol{p}_{1r}' + \text{contributions from other members meeting at joint } I, \\ \boldsymbol{p}_J &= \boldsymbol{p}_{2r}' + \text{contributions from other members meeting at joint } J, \end{aligned}\right\} \quad (5.10)$$

where the contributions from the other members will be $\boldsymbol{p}_1'$ vectors for those members with end 1 at the joint considered, and $\boldsymbol{p}_2'$ vectors for members with end 2 at the joint considered. Thus the load/displacement equations for joints I and J have the form

$$\left.\begin{aligned} \boldsymbol{p}_I &= (\boldsymbol{K}_{11}')_r\boldsymbol{d}_I + (\boldsymbol{K}_{12}')_r\boldsymbol{d}_J + \text{contributions from other members}, \\ \boldsymbol{p}_J &= (\boldsymbol{K}_{21}')_r\boldsymbol{d}_I + (\boldsymbol{K}_{22}')_r\boldsymbol{d}_J + \text{contributions from other members}. \end{aligned}\right\} \quad (5.11)$$

In this analysis we have assumed that $\boldsymbol{p}_I$ and $\boldsymbol{p}_J$ already incorporate any equivalent loads arising from distributed loads, etc., on member r and the other members meeting at joints I and J. If this is not so then equation (5.9a) must be replaced by

$$\left.\begin{aligned} \boldsymbol{p}_{1r}' &= (\boldsymbol{K}_{11}')_r\boldsymbol{d}_I + (\boldsymbol{K}_{12}')_r\boldsymbol{d}_J - (\boldsymbol{p}'_{\text{equiv}})_{1r}, \\ \boldsymbol{p}_{2r}' &= (\boldsymbol{K}_{21}')_r\boldsymbol{d}_I + (\boldsymbol{K}_{22}')_r\boldsymbol{d}_J - (\boldsymbol{p}'_{\text{equiv}})_{2r}. \end{aligned}\right\} \quad (5.9\text{b})$$

It is apparent that if (5.9b) is used to substitute for $\boldsymbol{p}_{1r}'$ and $\boldsymbol{p}_{2r}'$ in (5.10) then $(\boldsymbol{p}'_{\text{equiv}})_{1r}$ and $(\boldsymbol{p}'_{\text{equiv}})_{2r}$ will automatically be added to $\boldsymbol{p}_I$ and $\boldsymbol{p}_J$.

Thus it follows that the presence of a member between joints I and J has the effect of adding its $\boldsymbol{K}_{ij}'$ matrices to the stiffness matrix of the complete structure according to the following scheme:

$$\begin{array}{r} \\ \\ \\ \text{Row } I \\ \\ \\ \text{Row } J \\ \\ \\ \end{array} \begin{array}{c} \qquad\quad \text{Column } I \qquad\qquad\qquad\quad \text{Column } J \\ \left[\begin{array}{ccccc} & \cdot & & \cdot & \\ & \cdot & & \cdot & \\ & \cdot & & \cdot & \\ \cdots & \boldsymbol{K}_{11}' & \cdots\cdots\cdots & \boldsymbol{K}_{12}' & \cdots \\ & \cdot & & \cdot & \\ & \cdot & & \cdot & \\ \cdots & \boldsymbol{K}_{21}' & \cdots\cdots\cdots & \boldsymbol{K}_{22}' & \cdots \\ & \cdot & & \cdot & \\ & \cdot & & \cdot & \\ & \cdot & & \cdot & \end{array}\right] \end{array}$$

Since there is no load/displacement equation for the equilibrium of the fixed "joint" O, the final stiffness matrix **K** has no row or column corresponding to that joint. Thus a member whose end 1 is a point of foundation attachment merely contributes its K_{22}' matrix to the leading diagonal term of row J. Since K_{21}' is the transpose of K_{12}' the contributions of the individual members to the final matrix are always such as to maintain its symmetry, so that in practice we need only consider the assembly of the leading diagonal coefficients and those above them—the "upper triangle", as it is usually called. The rules for the assembly of this upper triangle may be stated as follows:

1. The leading diagonal sub-matrix in row I is the sum of the matrices K_{11}' or K_{22}' of all the members meeting at joint I, the matrix K_{11}' being selected if the member has end 1 at the joint, and the matrix K_{22}' being chosen if the end at the joint is end 2.
2. The off-diagonal sub-matrices in row I, corresponding to columns $J, K, L. \ldots > I$, are the K_{12}' matrices of the members connecting joint I to joints $J, K, L, \ldots$. If a joint is not directly connected to I by a member, then the associated sub-matrix is zero. (Due to the symmetry of the complete matrix it is only necessary to consider members which have end 1 at joint I.)

As an example, applying the rules given above to the frame shown in Fig. 5.4 produces the equations

$$\begin{bmatrix} p_A \\ \\ \\ p_B \\ \\ \\ p_C \\ \\ p_D \\ \end{bmatrix} = \begin{bmatrix} (K_{22}')_a & (K_{12}')_c & (K_{12}')_d & 0 \\ +(K_{11}')_c & & & \\ +(K_{11}')_d & & & \\ * & (K_{22}')_b & 0 & (K_{12}')_e \\ & +(K_{22}')_c & & \\ & +(K_{11}')_e & & \\ * & * & (K_{22}')_d & (K_{12}')_f \\ & & +(K_{11}')_f & \\ * & * & * & (K_{22}')_e \\ & & & +(K_{22}')_f \end{bmatrix} \begin{bmatrix} d_A \\ \\ \\ d_B \\ \\ \\ d_C \\ \\ d_D \\ \end{bmatrix}$$

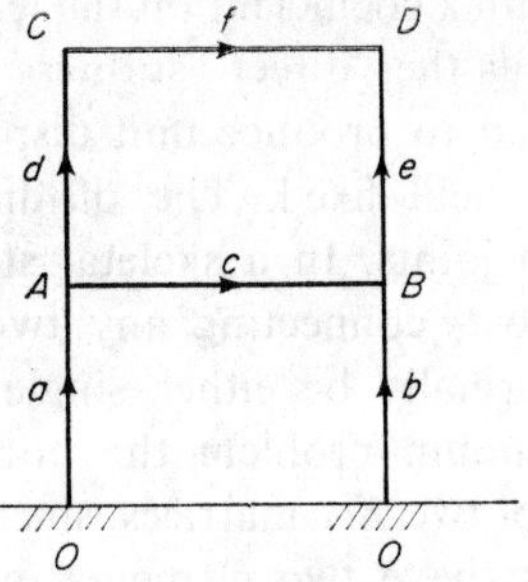

FIG. 5.4. Example of a rigid-jointed frame.

where only the upper triangle of the stiffness matrix has been shown.

Exactly the same procedure may be adopted in the assembly of the stiffness matrix **K** for a continuum problem. For example, consider a triangular element whose nodes (numbered 1, 2, 3 when we consider the triangle in isolation) are situated at nodes *I, J, K* of the complete finite element assembly. The nodal load/displacement equations are

$$\boldsymbol{p}_i' = \boldsymbol{K}_{ij}'\boldsymbol{d}_j' \quad (i, j = 1,2,3)$$

For a plane element the procedure for deriving the 2 × 2 matrices $\boldsymbol{K}_{ij}'$ is given in Section 4.2. The various $\boldsymbol{K}_{ij}'$ matrices are added to the overall stiffness matrix according to the scheme

$$\begin{array}{r} \\ \text{Row } I \\ \\ \text{Row } J \\ \\ \text{Row } K \\ \\ \end{array}
\begin{array}{c} \text{Column } I \qquad\qquad \text{Column } J \qquad\qquad \text{Column } K \\
\left[\begin{array}{ccccccc}
 & \cdot & & \cdot & & \cdot & \\
\cdot & \boldsymbol{K}_{11}' & \cdots & \boldsymbol{K}_{12}' & \cdots & \boldsymbol{K}_{13}' & \cdots \\
 & \cdot & & \cdot & & \cdot & \\
\cdot & \boldsymbol{K}_{21}' & \cdots & \boldsymbol{K}_{22}' & \cdots & \boldsymbol{K}_{23}' & \cdots \\
 & \cdot & & \cdot & & \cdot & \\
\cdot & \boldsymbol{K}_{31}' & \cdots & \boldsymbol{K}_{32}' & \cdots & \boldsymbol{K}_{33}' & \cdots \\
 & \cdot & & \cdot & & \cdot & \\
 & \cdot & & \cdot & & \cdot &
\end{array}\right]\end{array}$$

In any type of structure a coefficient on the leading diagonal of the final stiffness matrix represents the "direct" stiffness of a joint or node, i.e. the load required at the joint to produce unit displacement of that joint, all the other joints being held fixed. The off-diagonal terms correspond to connections between joints. In a skeletal structure there will usually be one member, at most, connecting any two joints, so that the off-diagonal terms will normally be either single $\boldsymbol{K}_{12}'$ or $\boldsymbol{K}_{21}'$ matrices or zero. In a plane continuum problem the non-zero off-diagonal terms will usually be the sum of two $\boldsymbol{K}_{ij}'$ matrices, since adjacent nodes not on an external boundary must have two elements in common. Similarly in a three-dimensional continuum problem adjacent internal nodes will have at least three elements in common.

In most skeletal structures joints are only connected to their nearer neighbours, and there are rarely more than six members meeting at any one joint. This means that, provided a systematic method of lettering or numbering the joints is adopted, the difference in the joint numbers at the ends of members will be small compared with the total number of joints, and as a result most of the non-zero coefficients in the stiffness matrix of the structure will lie near the leading diagonal. Such a matrix is called a *banded* matrix. The same phenomenon arises in continuum problems. This characteristic is important in the numerical solution of the load/displacement equations on a digital computer, and we discuss its implications in Section 11.1.

5.4. Settlement and partial restraint

In the previous sections we assumed that any member attached to a rigid foundation has all its components of displacement at the point of attachment equal to zero. In such cases the foundation joint is omitted from the analysis entirely. We now consider problems in which some of the displacement components at a foundation joint are non-zero.

We begin by considering foundation settlement. Imagine that the joint S shown in Fig. 5.5 has settled by an amount Δ, so that we need to include the displacement condition

$$\boldsymbol{d}_S = \begin{bmatrix} 0 \\ \Delta \\ 0 \end{bmatrix}$$

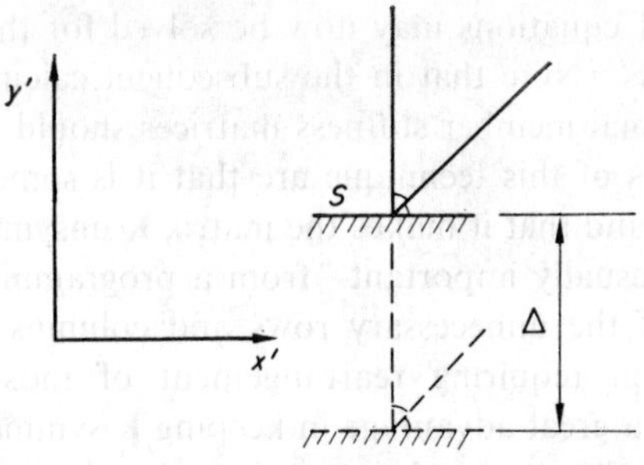

FIG. 5.5. Foundation settlement in a rigid-jointed framework.

in the load/displacement equations. If we treat S as a normal joint the process described in the previous section may be used to construct the complete set of equations

$$\begin{array}{l} \\ \\ \text{equilibrium} \\ \text{equations} \\ \text{for joint } S \\ \\ \\ \end{array}\left.\begin{array}{l} \\ \\ \\ \\ \\ \end{array}\right\} \begin{bmatrix} \cdot \\ \cdot \\ p_{xS}' \\ p_{yS}' \\ m_S' \\ \cdot \\ \cdot \\ \cdot \end{bmatrix} = \begin{bmatrix} & & & & & & & & & & \\ & & & & & & & & & & \\ \cdot & \cdot & \cdot & \cdot & \cdot & \cdot & \cdot & \cdot & \cdot & \cdot & \cdot \\ \cdot & \cdot & \cdot & \cdot & \cdot & \cdot & \cdot & \cdot & \cdot & \cdot & \cdot \\ \cdot & \cdot & \cdot & \cdot & \cdot & \cdot & \cdot & \cdot & \cdot & \cdot & \cdot \\ & & & & & & & & & & \\ & & & & & & & & & & \\ & & & & & & & & & & \end{bmatrix} \begin{bmatrix} \cdot \\ \cdot \\ \delta_{xS}' \\ \delta_{yS}' \\ \theta_S' \\ \cdot \\ \cdot \\ \cdot \end{bmatrix}$$

It is apparent that the equilibrium equations for joint S are not of direct use to us, since the reactions p_{xS}', p_{yS}', m_S' are unknown. We therefore replace them by equations which give the required values for the three components of displacement,

$$\begin{bmatrix} \cdot \\ \cdot \\ \cdot \\ 0 \\ \Delta \\ 0 \\ \cdot \\ \cdot \\ \cdot \end{bmatrix} = \begin{bmatrix} & & & & \\ & & & & \\ & & & & \\ 0 & \cdot\;\cdot\;\cdot\;\cdot\;\cdot & 0\;1\;0 & \cdot\;\cdot\;\cdot\;\cdot\;\cdot\;\cdot\;\cdot & 0 \\ 0 & \cdot\;\cdot\;\cdot\;\cdot\;\cdot\;\cdot & 0\;1\;0 & \cdot\;\cdot\;\cdot\;\cdot\;\cdot\;\cdot & 0 \\ 0 & \cdot\;\cdot\;\cdot\;\cdot\;\cdot\;\cdot\;\cdot & 0\;1\;0 & \cdot\;\cdot\;\cdot\;\cdot\;\cdot & 0 \\ & & & & \\ & & & & \\ & & & & \end{bmatrix} \begin{bmatrix} \cdot \\ \cdot \\ \cdot \\ \delta_{xS}' \\ \delta_{S}' \\ \theta_S' \\ \cdot \\ \cdot \\ \cdot \end{bmatrix}$$

[redacted]

This revised set of equations may now be solved for the complete set of joint displacements. (Note that in the subsequent calculation of member end-loads the normal member stiffness matrices should be used.)

The only defects of this technique are that it is somewhat wasteful of computer storage and that it makes the matrix **K** unsymmetric. The waste of storage is not usually important—from a programming point of view the elimination of the unnecessary rows and columns of **K** may be an awkward operation requiring rearrangement of most of the matrix. However, there is a great advantage in keeping **K** symmetric, since it may then be stored in upper (or lower) triangular form. In this example symmetry may easily be restored by adding multiples of the three altered equations to the other equations in such a way as to reduce the coefficients of δ_{xS}', δ_{yS}' and θ_S' in those equations to zero. It is apparent that this process only affects the loading vector **p** and the columns associated with δ_{xS}', δ_{yS}' and θ_S'.

The same idea may be used in the case of a joint which is pinned to a foundation. In this problem the moment equilibrium equation is similar to that of any other unrestrained joint and may be left unaltered. The equations in the x' and y' directions, however, involve unknown reactions and must be replaced by the equations $\delta_{xS}' = 0$, $\delta_{yS}' = 0$ (or $\delta_{xS}' = \Delta_x$, $\delta_{yS}' = \Delta_y$ if there is settlement). Overall symmetry may be restored by the technique described above.

A somewhat more complex example is shown in Fig. 5.6. If the angle β is zero we simply replace the equilibrium equation in the y' direction by the equation $\delta_{yS}' = 0$. If it is non-zero the condition which must be satisfied is $\delta_{yS}'' = -\delta_{xS}' \sin \beta + \delta_{yS}' \cos \beta = 0$. The simplest way of satisfying

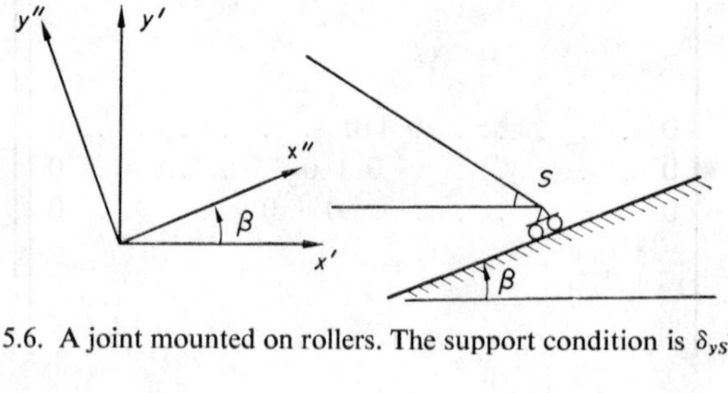

FIG. 5.6. A joint mounted on rollers. The support condition is $\delta_{yS}'' = 0$.

this condition is to change the load and displacement components p_{xS}', p_{yS}', δ_{xS}' and δ_{yS}' into the coordinate system x'', y''. This may be done by pre-multiplying the two rows of the assembled stiffness matrix associated with p_{xS}' and p_{yS}' by $\boldsymbol{T}_\beta$ and post-multiplying the two columns associated with δ_{xS}' and δ_{yS}' by $\boldsymbol{T}_\beta^t$, where

$$\boldsymbol{T}_\beta = \begin{bmatrix} \cos\beta & -\sin\beta \\ \sin\beta & \cos\beta \end{bmatrix}$$

This gives the set of equations

$$\begin{bmatrix} \cdot \\ \cdot \\ \cdot \\ \cdot \\ \cdot \\ p_{xS}'' \\ p_{yS}'' \\ \cdot \\ \cdot \\ \cdot \end{bmatrix} = \begin{bmatrix} & & & \cdot\ \cdot & & & \\ & & & \cdot\ \cdot & & & \\ & & & \cdot\ \cdot & & & \\ & & & \cdot\ \cdot & & & \\ & & & \cdot\ \cdot & & & \\ \cdot & \cdot & \cdot & \cdot\ \cdot & \cdot & \cdot & \cdot \\ \cdot & \cdot & \cdot & \cdot\ \cdot & \cdot & \cdot & \cdot \\ & & & \cdot\ \cdot & & & \\ & & & \cdot\ \cdot & & & \\ & & & \cdot\ \cdot & & & \end{bmatrix} \begin{bmatrix} \cdot \\ \cdot \\ \cdot \\ \cdot \\ \cdot \\ \delta_{xS}'' \\ \delta_{yS}'' \\ \cdot \\ \cdot \\ \cdot \end{bmatrix}$$

which are still symmetric. We now set $p_{xS}'' = 0$ and replace the equation for p_{yS}'' by the equation $\delta_{yS}'' = 0$. Overall symmetry may be restored by the same method as before and the equations solved for the joint displacements. (Note that it will be necessary to transform δ_{xS}'' and δ_{yS}'' back into the global coordinate system x', y' before the end-loads for the members meeting at S are computed.)

The treatment of a foundation as completely rigid is, of course, always an approximation. What we really mean in such cases is that the stiffness of the foundation is large compared with the stiffnesses of the members attached to it. If we wish we can include the elasticity of a foundation by treating each foundation joint as a "normal" joint attached by translational or rotational springs to points which do not displace. The stiffnesses of these equivalent springs are then added to the appropriate leading diagonal coefficients of the complete stiffness matrix **K**.

A problem of a slightly different sort is shown in Fig. 5.7, where we have a plane frame with two rigid joints connected by a pin. While there

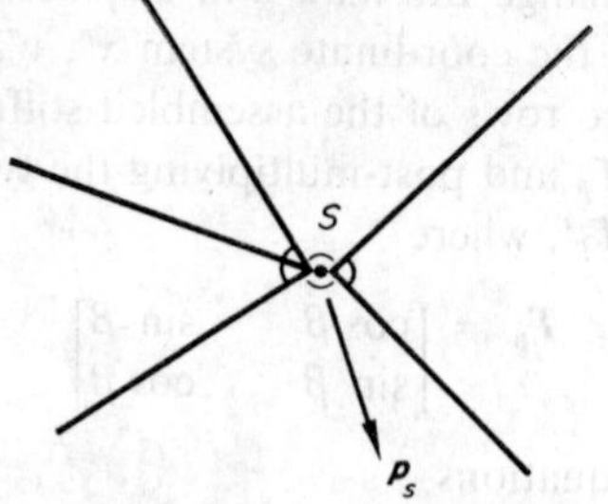

FIG. 5.7. A pin joint in a rigid-jointed frame.

are several ways of dealing with this situation the neatest is simply to give the joint an extra rotational degree of freedom. Thus the joint has the usual two translational components of displacement δ_{xS}' and δ_{yS}', a "left rotation" θ_{LS}' and a "right rotation" θ_{RS}'. The corresponding components of joint load are the force components p_{xS}', p_{yS}' and the left and right moments m_{LS}' and m_{RS}'. There are thus four rows and four columns of **K** associated with joint S, rather than the usual three. A member connected to joint S must be connected to either the left or right side of the pin. If connected to the left side its rotation at S is θ_{LS}' and it contributes its end-moment to m_{LS}'. The coefficients of its $\boldsymbol{K}'$ matrices are therefore added to the stiffness matrix of the structure according to the scheme shown in Fig. 5.8.

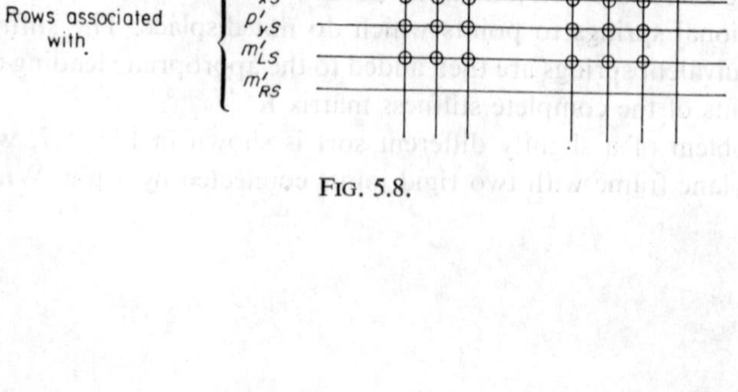

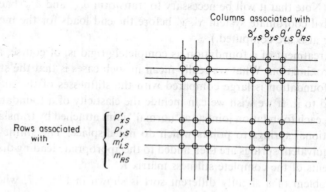

FIG. 5.8.

In the same way a member connected to the right-hand side of the pin contributes to the stiffness matrix of the structure as follows:

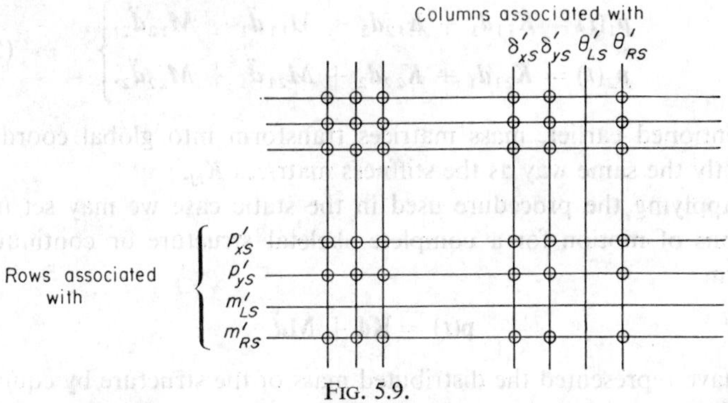

FIG. 5.9.

Techniques similar to those described in this section may be employed in cases where symmetry allows a reduction in the number of displacement variables. In all such problems the choice of approach depends a great deal on the form of computer storage used for the stiffness matrix **K**.

5.5. Dynamic problems

There are many instances where the response of a structure to time-dependent forces is required. For example, in designing a structure to resist earthquake or blast loading it is important to know the behaviour under transient loading conditions, while in the case of a structure supporting rotating machinery the natural frequencies of vibration are obviously of considerable importance. Problems of dynamic loading are particularly common in the field of aeronautics, and it is the complexity of these problems which has been largely responsible for the widespread adoption of matrix methods in the analysis of aircraft structures.

The equations of motion of a general undamped structural element under time-dependent nodal loads are

$$\boldsymbol{p}_i(t) = \boldsymbol{K}_{ij}\boldsymbol{d}_j + \boldsymbol{M}_{ij}\ddot{\boldsymbol{d}}_j \tag{5.12a}$$

where the mass matrices $\boldsymbol{M}_{ij}$ are defined in Section 3.11 (for line elements) and Section 4.6 (for area and volume elements). For a line element the equations may be written as

$$\left.\begin{aligned} \boldsymbol{p}_1(t) &= \boldsymbol{K}_{11}\boldsymbol{d}_1 + \boldsymbol{K}_{12}\boldsymbol{d}_2 + \boldsymbol{M}_{11}\ddot{\boldsymbol{d}}_1 + \boldsymbol{M}_{12}\ddot{\boldsymbol{d}}_2, \\ \boldsymbol{p}_2(t) &= \boldsymbol{K}_{21}\boldsymbol{d}_1 + \boldsymbol{K}_{22}\boldsymbol{d}_2 + \boldsymbol{M}_{21}\ddot{\boldsymbol{d}}_1 + \boldsymbol{M}_{22}\ddot{\boldsymbol{d}}_2. \end{aligned}\right\} \tag{5.12b}$$

As mentioned earlier, mass matrices transform into global coordinates in exactly the same way as the stiffness matrices $\boldsymbol{K}_{ij}$.

By applying the procedure used in the static case we may set up the equations of motion for a complete skeletal structure or continuum in the form

$$\mathbf{p}(t) = \mathbf{Kd} + \mathbf{M}\ddot{\boldsymbol{d}}. \tag{5.13}$$

If we have represented the distributed mass of the structure by equivalent lumped masses at the nodes then the mass matrix **M** will be a diagonal matrix. However, if we have used the consistent mass matrices developed in Sections 3.11 and 4.6 then **M** will have the same arrangement of zero and non-zero coefficients as **K**. In general (5.13) represents a set of linear differential equations, which may be integrated numerically from given initial conditions for any given loading $\mathbf{p}(t)$.

The commonest dynamic problem is the determination of natural frequencies and normal modes of free vibration. If the loading $\mathbf{p}(t)$ is zero and the displacement $\mathbf{d}$ has the form $\mathbf{d} = \mathbf{d}_0 \sin \omega t$ then (5.13) becomes

$$(\mathbf{K} - \omega^2\mathbf{M})\mathbf{d}_0 = \mathbf{0} \tag{5.14a}$$

or equivalently

$$|\mathbf{K} - \omega^2\mathbf{M}| = 0. \tag{5.14b}$$

Writing $\omega^2 = \lambda$ we obtain the classical eigenvalue problem of matrix algebra. Techniques for solving this problem are described in Section 5 of the Appendix.

Trouble may arise with certain types of numerical method if (5.14b) has zero or infinite roots. The root $\lambda = 0$ occurs when **K** is singular, i.e. when the structure has a rigid-body degree of freedom. The root $\lambda = \infty$ occurs when **M** is singular, i.e. when a displacement degree of freedom has no mass associated with it. This will happen with joint rotations in the lumped mass approach, and with both methods if certain parts of a structure are treated as weightless.

As an example of both types of mass representation we consider the uniform cantilever shown in Fig. 5.10. For *longitudinal* vibration the exact solution of the partial differential equation of motion gives the two lowest natural frequencies as

$$\omega_1 = \frac{\pi}{2L}\sqrt{\frac{E}{\rho}}, \quad \omega_2 = \frac{3\pi}{2L}\sqrt{\frac{E}{\rho}}.$$

Treating the beam as a single element gives a system with only one degree of freedom and hence only one natural frequency. Using the "lumped mass" method (Fig. 5.10b, equivalent mass of $\rho AL/2$ at the free end) we obtain $\omega = (\sqrt{2}/L)\sqrt{E/\rho}$, which is approximately 10% below the true value of ω_1. Using the "consistent mass matrix" method (Fig. 5.10a, equivalent mass of $\rho AL/3$ at the free end) we obtain $\omega = (\sqrt{3}/L)\sqrt{E/\rho}$, which is approximately 10% too high.

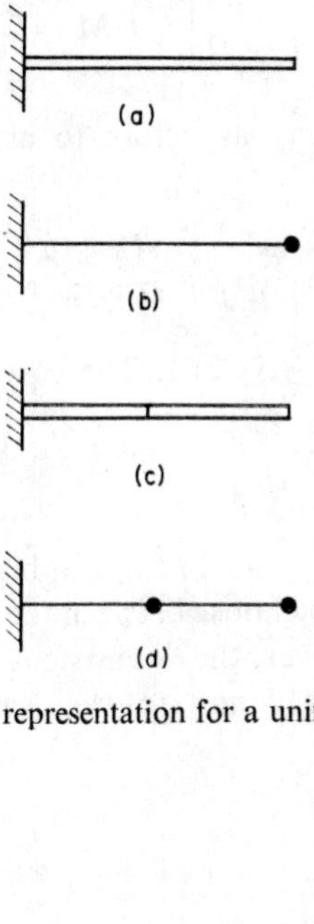

FIG. 5.10. Mass representation for a uniform cantilever.

If we split the beam into two elements the approximate analyses are as follows:

(a) Lumped mass approach (Fig. 5.10d): displacements δ_{x1}, δ_{x2}.

$$\mathbf{K} = \frac{2AE}{L}\begin{bmatrix} 2 & -1 \\ -1 & 1 \end{bmatrix} \qquad \mathbf{M} = \frac{\rho AL}{4}\begin{bmatrix} 2 & 0 \\ 0 & 1 \end{bmatrix}$$

If we write $\mu^2 = \rho\omega^2 L^2/8E$ equation (5.14b) becomes

$$\begin{vmatrix} 2 - 2\mu^2 & -1 \\ -1 & 1 - \mu^2 \end{vmatrix} = 0$$

which has roots $\mu^2 = 1 \pm 1/\sqrt{2}$. Hence we obtain natural frequencies

$$\omega_1 = \frac{1.531}{L}\sqrt{\frac{E}{\rho}}, \qquad \omega_2 = \frac{3.696}{L}\sqrt{\frac{E}{\rho}},$$

which are respectively 2.6% and 21.5% too low.

(b) Consistent mass matrix approach (Fig. 5.10c): displacements δ_{x1}, δ_{x2}.

$$\mathbf{K} = \frac{2AE}{L}\begin{bmatrix} 2 & -1 \\ -1 & 1 \end{bmatrix} \qquad \mathbf{M} = \frac{\rho AL}{12}\begin{bmatrix} 4 & 1 \\ 1 & 2 \end{bmatrix}$$

Defining μ in the same way as before to assist comparison, equation (5.14b) becomes

$$\begin{vmatrix} 6 - 4\mu^2 & -(3 + \mu^2) \\ -(3 + \mu^2) & 3 - 2\mu^2 \end{vmatrix} = 0$$

which has roots $\mu^2 = 3(5 \pm 3\sqrt{2})/7$. The corresponding frequencies are

$$\omega_1 = \frac{1.611}{L}\sqrt{\frac{E}{\rho}}, \qquad \omega_2 = \frac{5.629}{L}\sqrt{\frac{E}{\rho}},$$

which are respectively 2.6% and 19.5% too high.

So far there seems little to choose between the two methods of representing distributed mass. However, the "consistent mass matrix" method has a clear advantage when it comes to *transverse* vibrations. The exact value of the lowest natural frequency of transverse vibration for a uniform cantilever is

$$\omega_1 = \frac{3.515}{L^2}\sqrt{\frac{EI}{\rho A}}.$$

The approximate analyses are as follows,

(a) Lumped mass approach (Fig. 5.10d): two nodes, displacement components at each node δ_y, θ.

The **K** and **M** matrices are

$$\mathbf{K} = \frac{8EI}{L^3}\begin{bmatrix} 24 & 0 & -12 & 6L \\ 0 & 8L^2 & -6L & 2L^2 \\ -12 & 6L & 12 & -6L \\ 6L & 2L^2 & -6L & 4L^2 \end{bmatrix}, \quad \mathbf{M} = \frac{\rho AL}{4}\begin{bmatrix} 2 & 0 & 0 & 0 \\ 0 & 0 & 0 & 0 \\ 0 & 0 & 1 & 0 \\ 0 & 0 & 0 & 0 \end{bmatrix}$$

If we write $\mu^2 = \rho AL^4\omega^2/32EI$, equation (5.14b) becomes

$$\begin{vmatrix} 24-2\mu^2 & 0 & -12 & 6 \\ 0 & 8 & -6 & 2 \\ -12 & -6 & 12-\mu^2 & -6 \\ 6 & 2 & -6 & 4 \end{vmatrix} = 0$$

which reduces to $7\mu^4 - 60\mu^2 + 18 = 0$, the smallest root of which is $\mu = 0.5579$. The corresponding natural frequency is

$$\omega_1 = \frac{3.156}{L^2}\sqrt{\frac{EI}{\rho A}}$$

which is 10% too low.

(b) Consistent mass matrix approach (Fig. 5.10a): one node, displacement components δ_y, θ.

Treating the beam as a single element we have

$$\mathbf{K} = \frac{EI}{L^3}\begin{bmatrix} 12 & -6L \\ -6L & 4L^2 \end{bmatrix}, \quad \mathbf{M} = \frac{\rho AL}{420}\begin{bmatrix} 156 & -22L \\ -22L & 4L^2 \end{bmatrix}$$

With the same definition of μ^2 as before, equation (5.14b) becomes

$$\begin{vmatrix} 630 - 624\,\mu^2 & -315 + 88\,\mu^2 \\ -315 + 88\,\mu^2 & 210 - 16\,\mu^2 \end{vmatrix} = 0$$

which reduces to $448\,\mu^4 - 17{,}136\,\mu^2 + 6615 = 0$. The smallest root of this equation is $\mu = 0.6245$, which corresponds to a natural frequency

$$\omega_1 = \frac{3.533}{L^2}\sqrt{\frac{EI}{\rho A}} \text{ — an error of approximately } 0.5\%.$$

The use of consistent mass matrices always gives a value of the lowest natural frequency which is too high, as we have seen in these examples. This is because the technique considers the true mass distribution, but imposes a deformation mode which is not strictly correct at the natural frequency. It may indeed be regarded as a Rayleigh–Ritz approach in which the nodal displacements are the unknown parameters in the displacement function. The lumped mass approach, on the other hand, merely replaces one mass distribution by another which is similar as far as translational inertia is concerned. Its failure to represent rotational inertia means that beams in transverse vibration can only be represented adequately by the addition of a number of extra nodes along their length. From a computational point of view this is clearly an inefficient and undesirable procedure.

If it is only required to find the lowest natural frequency of a structure, equation (5.14a) may be written in the form $(\mathbf{M} - \lambda\mathbf{K})\mathbf{d}_0 = \mathbf{0}$, where $\lambda = 1/\omega^2$, and this may be transformed into $(\mathbf{A} - \lambda\mathbf{I})\mathbf{x} = \mathbf{0}$ by the method outlined in Section 5 of the Appendix, provided that $\mathbf{K}$ is positive definite. The largest value of λ now corresponds to the smallest value of ω, and may be found by the iterative method described in that section. In most structural problems the frequency of the first harmonic is considerably greater than the frequency of the fundamental mode, so that the iterative process usually converges with satisfactory rapidity.

Rayleigh's principle provides another convenient method for obtaining the lowest natural frequency. As mentioned above, when used with the consistent mass matrix approach it will always give a result which is slightly too high. When used with a lumped mass approximation one cannot say on which side of the true value the computed value will lie.

While the use of stiffness matrices makes it easy to formulate a structural vibration problem in classical eigenvalue terms, the heavily banded nature of the **K** and **M** matrices may make it practicable to find the solutions of equation (5.14b) by actually evaluating the determinant $|\mathbf{K} - \omega^2\mathbf{M}|$ numerically for a series of values of ω, and finding the zeros by a standard root-finding process. This approach is particularly useful if it is required to investigate the vibration characteristics of a structure in a given frequency range. It is often applied in conjunction with the use of transfer matrices and is discussed further in Section 9.6.

CHAPTER 6

The Equilibrium Equations of a Complete Structure

The equilibrium method described in the previous chapter assembles the stiffness matrix **K** from the $\boldsymbol{K}_{ij}'$ matrices associated with the individual elements. This procedure can be applied to both statically determinate and hyperstatic structures, and indeed one cannot deduce to which class a structure belongs simply by inspecting **K**.

An alternative approach is to separate the formation of the equilibrium equations from the specification of material properties. Just as the loads acting on the joints of a structure may be collected into a load vector **p**, so the member stress-resultants may be collected to form a stress-resultant vector **r**. The equilibrium equations which relate the vectors **p** and **r** can be written in the form

$$\mathbf{p} = \mathbf{H}\mathbf{r} \tag{6.1}$$

where the matrix **H** is built up from the individual member equilibrium matrices in very much the same way as **K** is built up from the individual member stiffness matrices.

This approach has a number of advantages. In the first place it provides a formal procedure for the analysis of statically determinate structures. For such structures the matrix **H** is square and non-singular, so that equations (6.1) can be solved for the member stress-resultants, quite independently of the material properties of the structure. Although the analysis of determinate structures is often regarded as elementary, the practical problem of stressing a complex three-dimensional lattice tower with several hundred members can hardly be dismissed as trivial, even if the members are arranged in such a way as to make the structure determinate. It is desirable, therefore, to have a formal procedure for setting up the equations which can be used as the basis of a computer program.

If a structure is hyperstatic then **H** has more columns than rows, so that one cannot solve (6.1) for the stress-resultant vector **r**. However, even in this case the matrix **H** is still of central importance. It is shown in Section 6.1 that the nodal compatibility equations for a structure may be written in the form

$$\mathbf{e} = \mathbf{H}^t\mathbf{d} \tag{6.2}$$

where **e** represents the vector formed from the individual member deformations. Equations (6.1) and (6.2) may be combined with appropriate relationships between **r** and **e** to provide an alternative form of the equilibrium method. This form is presented in Section 6.2.

The separation of the equilibrium equations (6.1) and the compatibility equations (6.2) from the equations defining the material properties has important advantages, particularly when dealing with non-linear structures. For structures whose members can be idealized as rigid-plastic, equations (6.1) provide a formal approach to plastic collapse analysis and minimum-weight design. Methods of solving these problems are presented in Chapter 7. The analysis of more general types of non-linear behaviour is discussed in Chapter 10.

As stated above, one cannot obtain a complete solution of the equilibrium equations (6.1) if a structure is hyperstatic. However, it is always possible to carry out a "partial solution", expressing **r** as a linear function of the applied loads **p** and an unknown vector **q** with fewer components than **r**,

$$\mathbf{r} = \mathbf{B}_0\mathbf{p} + \mathbf{B}\mathbf{q}. \tag{6.3}$$

This form of the equilibrium equations is the basis of the "compatibility" or "force" method of structural analysis. Equations (6.3) may be set-up by a direct physical argument or may be derived from (6.1) by purely algebraic means. Both approaches are described in Chapter 8.

6.1. Setting up the equilibrium equations

The analysis presented in this section is very similar to the procedure for the assembly of the load/displacement equations in the equilibrium method, as described in Sections 5.2 and 5.3.

We begin by considering the equilibrium of the truss shown in Fig. 6.1. We do not need to specify whether the truss has pinned or rigid joints, since the analysis holds equally for both, provided the relevant vectors and matrices are defined appropriately. Initially we assume that the joint loads $\boldsymbol{p}_A$, $\boldsymbol{p}_B$ include the effects of any distributed loads, etc., acting on the members.

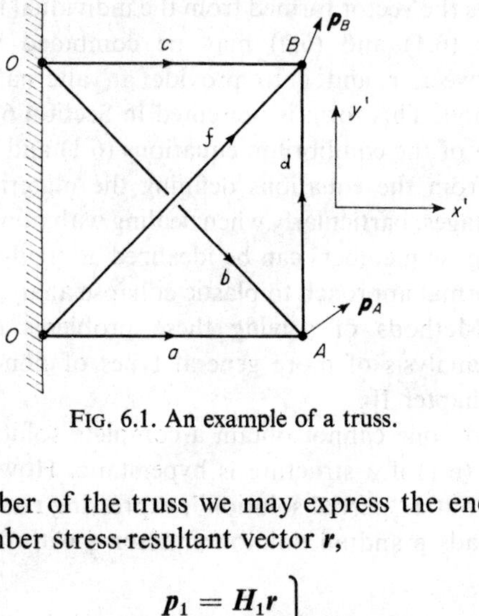

FIG. 6.1. An example of a truss.

For each member of the truss we may express the end-loads $\boldsymbol{p}_1$, $\boldsymbol{p}_2$ in terms of the member stress-resultant vector $\boldsymbol{r}$,

$$\left.\begin{aligned} \boldsymbol{p}_1 &= \boldsymbol{H}_1\boldsymbol{r} \\ \boldsymbol{p}_2 &= \boldsymbol{H}_2\boldsymbol{r} \end{aligned}\right\} \qquad (6.4a)$$

or, in the global coordinate system of the truss,

$$\left.\begin{aligned} \boldsymbol{p}_1' &= \boldsymbol{H}_1'\boldsymbol{r} \\ \boldsymbol{p}_2' &= \boldsymbol{H}_2'\boldsymbol{r} \end{aligned}\right\} \qquad (6.4b)$$

where $p_i' = Tp_i$ and $H_i' = TH_i$, $(i = 1,2)$.

The joint equilibrium equations are

$$\left.\begin{aligned} p_A &= p_{2a}' + p_{2b}' + p_{1d}', \\ p_B &= p_{2c}' + p_{2d}' + p_{2f}'. \end{aligned}\right\} \quad (6.5)$$

Substituting in these equations for p_{2a}', etc., from (6.4b) we obtain

$$p_A = H_{2a}'r_a + H_{2b}'r_b + H_{1d}'r_d,$$
$$p_B = H_{2c}'r_c + H_{2d}'r_d + H_{2f}'r_f.$$

These equations may be written in the form

$$\begin{bmatrix} p_A \\ p_B \end{bmatrix} = \begin{bmatrix} H_{2a}' & H_{2b}' & 0 & H_{1d}' & 0 \\ 0 & 0 & H_{2c}' & H_{2d}' & H_{2f}' \end{bmatrix} \begin{bmatrix} r_a \\ r_b \\ r_c \\ r_d \\ r_f \end{bmatrix} \quad (6.6a)$$

or as

$$\mathbf{p} = \mathbf{Hr}. \quad (6.6b)$$

We see from (6.6a) that each row of $\mathbf{H}$ corresponds to a joint, while each column corresponds to a member. The non-zero matrices H_{2a}', H_{2b}', H_{1d}' in the first row correspond to the members which meet at joint A, the subscript 1 or 2 depending on whether the member has end 1 or end 2 at the joint.

If the equivalent end-loads due to distributed loading on the members, etc., are not included in p_A and p_B then equations (6.4a) must be replaced by

$$\left.\begin{aligned} p_1 &= H_1 r - (p_{\text{equiv}})_1, \\ p_2 &= H_2 r - (p_{\text{equiv}})_2 \end{aligned}\right\} \quad (6.7a)$$

and (6.4b) by

$$\left.\begin{aligned} p_1' &= H_1' r - (p'_{\text{equiv}})_1, \\ p_2' &= H_2' r - (p'_{\text{equiv}})_2. \end{aligned}\right\} \quad (6.7b)$$

As with (5.9b), it is apparent that if (6.7b) is used to substitute for the various end-loads p_{2a}', etc., appearing in (6.5) then the equivalent loads will automatically be added to p_A and p_B.

It is a small step from the above example to the equations for a general frame. If we remember that each row of **H** corresponds to a joint and each column to a member it is apparent that the sub-matrix in row I and column j of **H** refers to joint I and member j. If member j has end 1 at joint I then the sub-matrix is H_{1j}'. If member j has end 2 at joint I then the sub-matrix is H_{2j}'. If neither end of the member is connected to the joint then the sub-matrix is a zero matrix of appropriate size. The reader may confirm that equation (6.6a) conforms to these rules. As with the stiffness matrix **K**, there is no row of **H** corresponding to points of attachment to a rigid foundation.

In the case of a continuum problem the procedure is similar. For example, a triangular element s whose vertices 1, 2, 3 are situated at nodes I, J, K of a mesh contributes H_{1s}' to row I, H_{2s}' to row J and H_{3s}' to row K of column s of **H**.

In the case of single elements we found that the equilibrium matrix H reappears transposed in the compatibility equations relating the element deformation vector e and the nodal displacements d. The assembled equilibrium matrix **H** reappears in a similar way in the complete set of nodal compatibility equations. If we combine the individual element deformation vectors into a vector **e** then the relation between **e** and the vector of joint displacements **d** is

$$\mathbf{e} = \mathbf{H}^t\mathbf{d}. \tag{6.8}$$

The proof follows familiar lines. If we let a set of applied loads **p** move through some set of virtual joint displacements **d*** then the virtual work equation is

$$\mathbf{p}^t\mathbf{d}^* = \mathbf{r}^t\mathbf{e}^*$$

where **e*** is the set of virtual deformations compatible with **d***. Substituting for $\mathbf{p}^t$ from (6.6b) we obtain

$$\mathbf{r}^t\mathbf{e}^* = \mathbf{r}^t\mathbf{H}^t\mathbf{d}^*$$

and since **r** is independent of the virtual displacements we have $\mathbf{e}^* = \mathbf{H}^t\mathbf{d}^*$.

We now argue that this relationship must also hold between real deformations **e** and real displacements **d**, provided these are sufficiently small for the equilibrium matrix **H** to remain constant during the deformation. As mentioned earlier, this corresponds to the normal procedure of writing down the equilibrium conditions using the geometry of the original undistorted structure.

6.2. An alternative form of the equilibrium method

The matrix **H** derived in the previous section provides a complete mathematical description of the way in which the members of a structure are connected together, including the orientation of the members to the global coordinate axes. However, it contains no information about the material properties of the elements.

To link equations (6.6b) and (6.8) we need to relate the vectors **r** and **e**. If the members are linear-elastic we have the relationship $\boldsymbol{r} = \boldsymbol{Ke}$ for each individual member, so that all we need to do is combine these individual equations to form a single matrix equation. For the example discussed at the beginning of the previous section this equation will be

$$\begin{bmatrix} \boldsymbol{r}_a \\ \boldsymbol{r}_b \\ \boldsymbol{r}_c \\ \boldsymbol{r}_d \\ \boldsymbol{r}_f \end{bmatrix} = \begin{bmatrix} \boldsymbol{K}_a & 0 & 0 & 0 & 0 \\ 0 & \boldsymbol{K}_b & 0 & 0 & 0 \\ 0 & 0 & \boldsymbol{K}_c & 0 & 0 \\ 0 & 0 & 0 & \boldsymbol{K}_d & 0 \\ 0 & 0 & 0 & 0 & \boldsymbol{K}_f \end{bmatrix} \begin{bmatrix} \boldsymbol{e}_a \\ \boldsymbol{e}_b \\ \boldsymbol{e}_c \\ \boldsymbol{e}_d \\ \boldsymbol{e}_f \end{bmatrix} \tag{6.9a}$$

As we saw in Chapters 3 and 4, the individual member $\boldsymbol{K}$ matrices are diagonal for many common elements, or can be made so by a suitable choice of components for $\boldsymbol{r}$ and $\boldsymbol{e}$. Equation (6.9a) may be written as

$$\mathbf{r} = \mathbf{K}_{\mathbf{m}}\mathbf{e} \tag{6.9b}$$

where the subscript **m** stands for "members". We may if we like regard $\mathbf{K}_{\mathbf{m}}$ as the stiffness matrix of the unassembled structure.

If we now combine (6.6b), (6.8) and (6.9b) we obtain

$$\mathbf{p} = \mathbf{Hr} = \mathbf{HK}_{\mathbf{m}}\mathbf{e} = \mathbf{HK}_{\mathbf{m}}\mathbf{H}^t\mathbf{d} \tag{6.10a}$$

which we can write in the familiar form

$$\mathbf{p} = \mathbf{K}\mathbf{d} \tag{6.10b}$$

where $\mathbf{K}$, the stiffness matrix of the assembled structure, is equal to $\mathbf{H}\mathbf{K}_m\mathbf{H}^t$. The parallel with the result $K_{ij} = H_i K H_j^t$ for the individual element stiffness matrices is obvious. The reader may verify that for the matrices $\mathbf{H}$ and $\mathbf{K}_m$ defined by (6.6a) and (6.9a) the product $\mathbf{H}\mathbf{K}_m\mathbf{H}^t$ is essentially the same as the matrix $\mathbf{K}$ which appears in equation (5.8). From a practical point of view it is easier to assemble $\mathbf{K}$ from the rules given in Section 5.3 than by the process described in this section. However, the expression for $\mathbf{K}$ given in (6.10) is the starting-point for a number of useful ideas, as we shall see in the remainder of this chapter.

6.3. The analysis of determinate structures

As mentioned earlier, a determinate structure is one in which the equations of equilibrium are sufficient to determine all the internal nodal loads, i.e. it is a structure for which the equilibrium matrix $\mathbf{H}$ is square and non-singular, so that (6.1) may be written in the form $\mathbf{r} = \mathbf{H}^{-1}\mathbf{p}$. In these circumstances (6.10a) may be written in the form

$$\mathbf{d} = (\mathbf{H}\mathbf{K}_m\mathbf{H}^t)^{-1}\mathbf{p} = (\mathbf{H}^t)^{-1}\mathbf{K}_m^{-1}\mathbf{H}^{-1}\mathbf{p}. \tag{6.11}$$

Since $\mathbf{K}_m$ is simply the diagonal matrix formed from the individual member stiffness matrices it follows that $\mathbf{K}_m^{-1}$ is also a diagonal matrix, consisting of the individual member flexibility matrices. Thus we write $\mathbf{K}_m^{-1} = \mathbf{F}_m$, and since $(\mathbf{H}^t)^{-1} = (\mathbf{H}^{-1})^t$ equation (6.11) becomes

$$\mathbf{d} = (\mathbf{H}^{-1})^t\mathbf{F}_m\mathbf{H}^{-1}\mathbf{p}. \tag{6.12}$$

The symbolism of (6.12) should not conceal the fact that this equation is simply a description of the normal way in which we find the displacements of a statically determinate structure. The process may be split into three steps:

1. The individual member stress-resultants are found in terms of the known external loads by solving the equations of joint equilibrium: $\mathbf{r} = \mathbf{H}^{-1}\mathbf{p}$.
2. The member deformations are found from the member stress-resultants and the elastic characteristics of the members: $\mathbf{e} = \mathbf{F}_m\mathbf{r}$.

3. The joint displacements are computed from the member deformations: $\mathbf{d} = (\mathbf{H}^{-1})^t\mathbf{e}$. Note that $\mathbf{H}^{-1}$ is already available from (1).

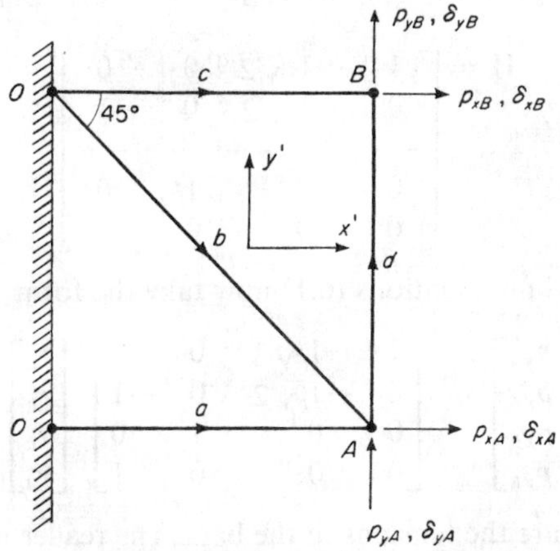

FIG. 6.2. An example of a determinate truss.

As an example of this procedure, consider the frame shown in Fig. 6.2. The rules set out in Section 6.1 give us

$$\mathbf{H} = \begin{bmatrix} H_{2a}' & H_{2b}' & 0 & H_{1d}' \\ 0 & 0 & H_{2c}' & H_{2d}' \end{bmatrix} \tag{6.13}$$

If the frame has rigid joints then each H' matrix is square. It follows that the matrix $\mathbf{H}$ in equation (6.13) is rectangular, so that we cannot solve (6.1) for $\mathbf{r}$ in terms of $\mathbf{p}$. The frame is therefore statically indeterminate.

Now consider the case where the structure has pinned joints. For a pin-ended member

$$T = \begin{bmatrix} \cos\alpha \\ \sin\alpha \end{bmatrix}$$

(see (3.19a)), while $H_1 = -1$, $H_2 = +1$.

$$\text{Thus } H_1' = -\begin{bmatrix} \cos\alpha \\ \sin\alpha \end{bmatrix}, \quad H_2' = \begin{bmatrix} \cos\alpha \\ \sin\alpha \end{bmatrix}$$

It follows from the figure that

$$\boldsymbol{H}_{2a}' = \boldsymbol{H}_{2c}' = \begin{bmatrix} 1 \\ 0 \end{bmatrix}, \quad \boldsymbol{H}_{2b}' = \begin{bmatrix} 1/\sqrt{2} \\ -1/\sqrt{2} \end{bmatrix}, \quad \boldsymbol{H}_{1d}' = -\begin{bmatrix} 0 \\ 1 \end{bmatrix}, \quad \boldsymbol{H}_{2d}' = \begin{bmatrix} 0 \\ 1 \end{bmatrix},$$

so that

$$\mathbf{H} = \left[\begin{array}{c|c|c|c} 1 & 1/\sqrt{2} & 0 & 0 \\ 0 & -1/\sqrt{2} & 0 & -1 \\ \hline 0 & 0 & 1 & 0 \\ 0 & 0 & 0 & 1 \end{array}\right]$$

The joint equilibrium equations (6.1) now take the form

$$\begin{bmatrix} p_{xA} \\ p_{yA} \\ p_{xB} \\ p_{yB} \end{bmatrix} = \begin{bmatrix} 1 & 1/\sqrt{2} & 0 & 0 \\ 0 & -1/\sqrt{2} & 0 & -1 \\ 0 & 0 & 1 & 0 \\ 0 & 0 & 0 & 1 \end{bmatrix} \begin{bmatrix} r_a \\ r_b \\ r_c \\ r_d \end{bmatrix} \tag{6.14}$$

where r_a, r_b, etc., are the tensions in the bars. The reader may easily verify that these equations are identical with the equations derived by a direct consideration of the forces in the truss. It will be seen that **H** is non-singular, as expected. Equations (6.14) may be inverted to give

$$\begin{bmatrix} r_a \\ r_b \\ r_c \\ r_d \end{bmatrix} = \begin{bmatrix} 1 & 1 & 0 & 1 \\ 0 & -\sqrt{2} & 0 & -\sqrt{2} \\ 0 & 0 & 1 & 0 \\ 0 & 0 & 0 & 1 \end{bmatrix} \begin{bmatrix} p_{xA} \\ p_{yA} \\ p_{xB} \\ p_{yB} \end{bmatrix} \tag{6.15}$$

The triangular form of **H** makes the inversion process extremely easy, and produces a matrix $\mathbf{H}^{-1}$ which is also an upper triangular matrix. In this example equation (6.15) may easily be checked by statics.

As we showed in Chapter 3, the "stiffness matrix" of a straight uniform pin-ended member is simply the single coefficient EA/L. The "flexibility matrix" is therefore simply L/EA. If the members of our truss are straight and of uniform cross-section then $\mathbf{F_m}$ has the form

$$\mathbf{F_m} = \begin{bmatrix} (L/EA)_a & 0 & 0 & 0 \\ 0 & (L/EA)_b & 0 & 0 \\ 0 & 0 & (L/EA)_c & 0 \\ 0 & 0 & 0 & (L/EA)_d \end{bmatrix}$$

Having obtained the matrices $\mathbf{H}^{-1}$ and $\mathbf{F}_m$ the rest of the analysis is merely a matter of carrying out the second and third of the steps listed above. Note that in this example the vector $\mathbf{e}$ is simply the vector of bar extensions.

While the simple truss which we have considered is determinate if its deformation is restricted to a plane it becomes a mechanism when considered as a three-dimensional structure. (The pin-joints at the ends of the members are assumed to be replaced by ball-joints.) This is reflected in the form of the $\mathbf{H}$ matrix, which becomes rectangular when the $\boldsymbol{H}'$ matrices appearing in it are those associated with a general pin-ended member in three dimensions. Each of these matrices consists of a column of three components—the general form may be obtained from (3.19c) as

$$H_1' = -\begin{bmatrix} l \\ m \\ n \end{bmatrix}, \qquad H_2' = \begin{bmatrix} l \\ m \\ n \end{bmatrix}$$

The matrix $\mathbf{H}$ has therefore more rows than columns, so that equation (6.1) has no general solution. There may of course be a solution for particular values of the external load vector $\mathbf{p}$, but this is not sufficient to make the assembly of members a "structure" in the usual sense. This form of $\mathbf{H}$ is characteristic of a mechanism.

Finally, it should be realized that the analysis described in this and the previous sections is not restricted to structures whose members are all of the same type. For example, consider the case where Fig. 6.2 represents a three-dimensional frame in which the members *a*, *c* and *d* are rigidly joined together at their ends, but the member *b* can only carry an axial force. In these circumstances the matrices H_{2a}', H_{2c}', H_{1d}' and H_{2d}' appearing in (6.13) are the appropriate square matrices associated with members whose ends can carry bending moments, while the matrix H_{2b}' consists of a single column. It is necessary to add zeros to this column in the rows corresponding to the moment components of p_A—it is clear that the axial force in member *b* can make no contribution to these components. The members *a*, *c* and *d* each contribute a deformation and stress-resultant vector containing six components to the vectors $\mathbf{e}$ and $\mathbf{r}$, while the member *b* only contributes a single term to each. The matrix $\mathbf{K}_m$ contains the appropriate square matrices for members *a*, *c* and *d*, and a single coefficient for the member *b*. With the various vectors and matrices defined in this way the formal solution provided by equations (6.10) and (6.11) still holds.

6.4. Determinate structures with rigid joints: tree structures

If we consider the rigid-jointed frame shown in Fig. 6.2 it is clear that a load such as $\boldsymbol{p}_A$ must eventually be transmitted to the foundations of the structure. In this particular example there are three alternative paths for the transmission of this load—the member *a*, the member *b*, and the pair of members *c* and *d*. Without information about the relative stiffness of these members we cannot tell what proportion of $\boldsymbol{p}_A$ will follow each path. We cannot therefore determine the internal forces and moments from statics alone, so that the structure is hyperstatic. From this argument it follows that for a rigid-jointed structure to be determinate there must be only one path to the foundations from any given loading point. This implies that the structure must only be attached to a rigid foundation at a single point, and that the members must be arranged in such a way that they do not form any closed rings. Such a structure we term a *tree structure*. At first sight the analysis of tree structures might appear to be merely an academic exercise, since few engineering structures are of this form. However, as we shall see in Chapter 8, the application of the compatibility method to a rigid-jointed structure involves reducing the structure to determinate form by the introduction of suitable "releases". If these releases take the form of complete discontinuities in the structure the result is a tree structure. It is for this reason that the problem is discussed here.

The equilibrium matrix $\mathbf{H}$ for a tree structure may be set up by following the rules given in Section 6.1. Since the structure is statically determinate the inverse matrix $\mathbf{H}^{-1}$ exists and may be found by ordinary numerical means. However, it is also possible to set up $\mathbf{H}^{-1}$ by direct physical reasoning. This is clearly more efficient from a computational point of view. As an example we consider the structure shown in Fig. 6.3, in which loads may be applied at any of the joints $A, B, \ldots, F$.

The first step is to express the flexibility matrix $\boldsymbol{F}$ and the matrices $\boldsymbol{H}_1'$, $\boldsymbol{H}_2'$ of each member in the coordinate system $Ox'y'$. (Flexibility matrices which have been computed in individual member coordinate

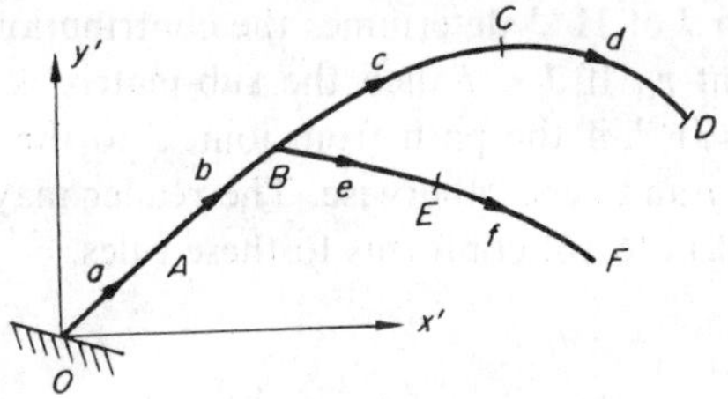

FIG. 6.3. A tree structure containing two branches.

systems may be transformed into this system by means of equation (3.30a).) In this coordinate system the vector $\boldsymbol{r}$ associated with each member is an equivalent force and moment at the origin O (see Section 3.1). Note that since all members are referred to a common origin $\boldsymbol{H}_{2a}' = -\boldsymbol{H}_{1b}'$, etc.

If we consider member b, the value of $\boldsymbol{r}_b$ depends on the loads at B, C, D, E and F. However, it is not affected by the load at A since the point A is between the root of the tree and the member. Similarly the value of $\boldsymbol{r}_c$ depends only on the loads at C and D. Using the notation of Section 3.1, the load $\boldsymbol{p}_F$ is equivalent to a load $(\boldsymbol{H}_{2f}')^{-1}\boldsymbol{p}_F$ at the origin, where

$$\boldsymbol{H}_{2f}' = \begin{bmatrix} 1 & 0 & 0 \\ 0 & 1 & 0 \\ y_F & -x_F & 1 \end{bmatrix}, \quad (\boldsymbol{H}_{2f}')^{-1} = \begin{bmatrix} 1 & 0 & 0 \\ 0 & 1 & 0 \\ -y_F & x_F & 1 \end{bmatrix}$$

for a plane structure. (The three-dimensional form is straightforward to construct.) It follows that

$$\boldsymbol{r}_b = (\boldsymbol{H}_{2b}')^{-1}\boldsymbol{p}_B + (\boldsymbol{H}_{2c}')^{-1}\boldsymbol{p}_C + (\boldsymbol{H}_{2d}')^{-1}\boldsymbol{p}_D + (\boldsymbol{H}_{2e}')^{-1}\boldsymbol{p}_E + (\boldsymbol{H}_{2f}')^{-1}\boldsymbol{p}_F.$$

Applying the same procedure to the other joints of the structure we obtain

$$\mathbf{H}^{-1} = \begin{bmatrix} (\boldsymbol{H}_{2a}')^{-1} & (\boldsymbol{H}_{2b}')^{-1} & (\boldsymbol{H}_{2c}')^{-1} & (\boldsymbol{H}_{2d}')^{-1} & (\boldsymbol{H}_{2e}')^{-1} & (\boldsymbol{H}_{2f}')^{-1} \\ \boldsymbol{0} & (\boldsymbol{H}_{2b}')^{-1} & (\boldsymbol{H}_{2c}')^{-1} & (\boldsymbol{H}_{2d}')^{-1} & (\boldsymbol{H}_{2e}')^{-1} & (\boldsymbol{H}_{2f}')^{-1} \\ \boldsymbol{0} & \boldsymbol{0} & (\boldsymbol{H}_{2c}')^{-1} & (\boldsymbol{H}_{2d}')^{-1} & \boldsymbol{0} & \boldsymbol{0} \\ \boldsymbol{0} & \boldsymbol{0} & \boldsymbol{0} & (\boldsymbol{H}_{2d}')^{-1} & \boldsymbol{0} & \boldsymbol{0} \\ \boldsymbol{0} & \boldsymbol{0} & \boldsymbol{0} & \boldsymbol{0} & (\boldsymbol{H}_{2e}')^{-1} & (\boldsymbol{H}_{2f}')^{-1} \\ \boldsymbol{0} & \boldsymbol{0} & \boldsymbol{0} & \boldsymbol{0} & \boldsymbol{0} & (\boldsymbol{H}_{2f}')^{-1} \end{bmatrix} \tag{6.16}$$

If we have a general tree structure and label it in the way indicated in Fig. 6.3 (that is, joint Q at end 2 of member q) then the sub-matrix in row i and column J of $\mathbf{H}^{-1}$ determines the contribution of the load $\boldsymbol{p}_J$ to the stress-resultant $\boldsymbol{r}_i$. If $J < i$ then the sub-matrix is zero. If $J \geqslant i$ the sub-matrix is $(\boldsymbol{H}_{2j}')^{-1}$ if the path from joint J to the foundation passes through member i and zero otherwise. The reader may easily verify that the matrix set out in (6.16) conforms to these rules.

CHAPTER 7

Plastic Analysis and Design

So far our analysis has been based on small-displacement elastic theory, with strains assumed proportional to stresses and displacements consequently proportional to loads. The assumptions of this theory, which are the traditional ones of structural engineering analysis, allow us to predict stresses and displacements under working loads with sufficient accuracy for design purposes. In contrast, the plastic theory of structures† ignores all elastic deformations and assumes that a structure does not distort at all until sufficient plastic regions have formed to convert it into a mechanism. Plastic theory is therefore concerned solely with structures at the point of collapse. It is sometimes referred to as *limit state* theory.

Historically, plastic methods of analysis and design were developed quite separately from elastic methods, and indeed there was a time when the two seemed to be in direct competition. However, plastic methods are similar to elastic methods in the sense that both generate well-known problems in linear algebra. The assumptions and approximations of *elastic* small-displacement theory lead, as we have seen, to the computational problems of solving a set of linear equations or inverting a matrix. With suitable assumptions, a systematic approach to a number of common problems in *plastic* small-displacement theory leads to the somewhat more complex problem of maximizing a linear function of a set of variables subject to a set of linear constraints—the "linear programming" problem. A formal statement of this problem and a discussion of the nature of its solution will be found in Section 7 of the Appendix.

† For a full account of the assumptions of plastic structural analysis and the manual methods of solving the problems discussed in this chapter the reader is referred to the standard texts by Neal (1963) and Baker and Heyman (1969).

The purpose of this chapter is to present a systematic treatment of those problems in plastic analysis and design which can be reduced to standard linear programming form, using the notation and ideas developed in previous chapters. The problems we shall discuss are all associated with plane skeletal frames and trusses whose members are initially straight, acted on by concentrated loads applied at specified points. The problems are:

1. The rigid-plastic collapse analysis of a frame or truss under proportional loading.
2. The shakedown analysis of a frame or truss. (This requires preliminary elastic analyses of the structure.)
3. Minimum-weight design of a frame or truss under a single known loading system.
4. Minimum-weight design for variable loads, but without including the possibility of incremental collapse.

Similar techniques may be used to set up the relevant equations and inequalities in continuum problems. However, the inequalities are often non-linear, so that different (and considerably more complicated) computational methods must be used.

In our discussion of elastic analysis we tacitly assumed that the structural engineer wishing to solve a set of linear equations or invert a matrix will use a standard computer routine which he need not necessarily understand. In the same way we may regard a problem in plastic analysis or design as "solved" once it has been reduced to standard linear programming form, since all computers have routines for this general algebraic problem. However, there are certain disadvantages in this attitude. As we shall see, a formal linear programming approach to problems in plasticity tends to generate large sparse matrices. The use of standard library programs may therefore be wasteful of computer time and storage unless special precautions are taken. The chapter concludes with an account of an algorithm for solving the first of the problems listed above which is considerably more compact than the standard procedure.

7.1. Ideal rigid-plastic behaviour

In this chapter we consider plane structures with initially straight members having the following characteristics:

(a) The tension t always lies between limits $-t^L$ and t^U. If $-t^L < t < t^U$ then the axial extension e is assumed to be zero. If $t = t^U$ then the axial extension may have any value $\geqslant 0$, while if $t = -t^L$ it may have any value $\leqslant 0$. The tension/extension graph is shown in Fig. 7.1. In practice the compressive limit $-t^L$ may be determined by buckling considerations rather than material yield, but this does not affect the remainder of the analysis. When applying axial yield criteria we assume that the tension in each member is constant along its length (i.e. any longitudinal loads applied at interior points of members are simply replaced by equivalent end-loads).

(b) The moment m at any cross-section always lies between $\pm m_p$, where m_p is termed the plastic moment of resistance and may vary along the length of the member. If $m = \pm m_p$ a plastic hinge is said to have formed, and the parts of the member on either side of the hinge can rotate relative to each other without change in the resisting moment, provided that the direction of rotation is such that positive work is done on the hinge. The moment/rotation graph is shown in Fig. 7.2.

In the analysis which follows, both types of constraint are always included. This means that the procedures we shall describe are equally applicable to frames (where the possibility of direct axial yield is usually remote) and trusses (where collapse requires some axial yield). In practice the value of the fully plastic moment m_p is a function of the axial force t. In the present analysis, however, we ignore this dependence and treat

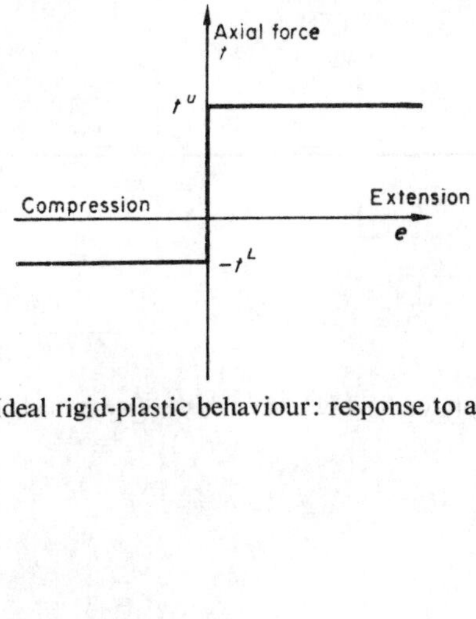

FIG. 7.1. Ideal rigid-plastic behaviour: response to axial force.

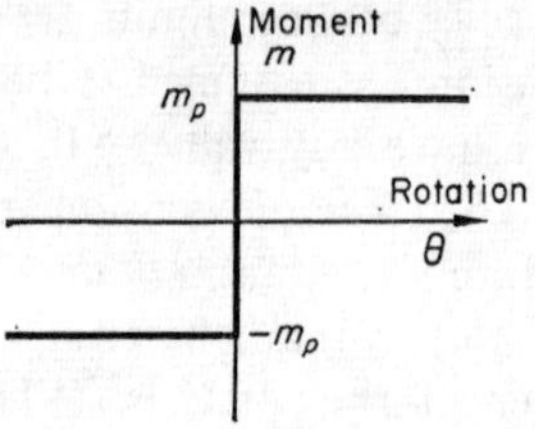

FIG. 7.2. Ideal rigid-plastic behaviour: behaviour at a plastic hinge.

t^L, t^U and m_p as constants dependent only on the member dimensions and the yield stress of the material.

7.2. Setting up the equilibrium equations

When we considered the analysis of a straight uniform elastic beam in Chapter 3 we chose the components of the stress-resultant vector $\boldsymbol{r}$ in a way which made the member stiffness matrix $\boldsymbol{K}$ diagonal. For plastic analysis we choose the components in a different way.

If a beam is loaded by concentrated transverse loads the greatest moment will always be either under a load or at one of the ends. This suggests that it is appropriate to include the end moments m_1, m_2 in the vector $\boldsymbol{r}$. In practice, however, plastic hinges cannot occur exactly at the ends of members, since real joints are always of finite size. We therefore choose for the components of $\boldsymbol{r}$ the axial tension t and the moments $\tilde{m}_1$, $\tilde{m}_2$ at a distance ϵL from the ends, as shown in Fig. 7.3.

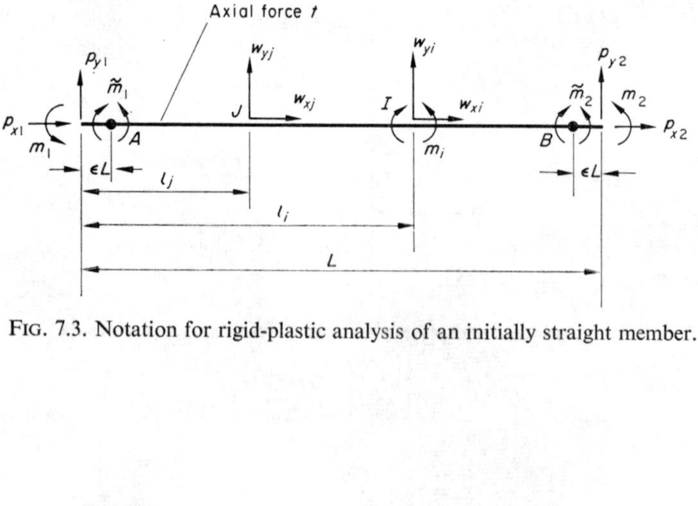

FIG. 7.3. Notation for rigid-plastic analysis of an initially straight member.

With $\boldsymbol{r}$ defined in this way the equilibrium equations for the member are

$$\left.\begin{aligned} \boldsymbol{p}_1 &= \boldsymbol{H}_1\boldsymbol{r} - (\boldsymbol{p}_{\text{equiv}})_1 \\ \boldsymbol{p}_2 &= \boldsymbol{H}_2\boldsymbol{r} - (\boldsymbol{p}_{\text{equiv}})_2 \end{aligned}\right\} \tag{7.1}$$

where

$$\boldsymbol{H}_1 = \begin{bmatrix} -1 & 0 & 0 \\ 0 & a & -a \\ 0 & 1+b & -b \end{bmatrix}, \quad \boldsymbol{H}_2 = \begin{bmatrix} 1 & 0 & 0 \\ 0 & -a & a \\ 0 & b & -(1+b) \end{bmatrix}, \quad \boldsymbol{r} = \begin{bmatrix} t \\ \tilde{m}_1 \\ \tilde{m}_2 \end{bmatrix}$$

with

$$a = \frac{1}{(1-2\epsilon)L}, \qquad b = \frac{\epsilon}{1-2\epsilon}$$

and

$$(\boldsymbol{p}_{\text{equiv}})_1 = \begin{bmatrix} \sum (L-l_i)W_{xi}/L \\ a\sum c_{1i}W_{yi} \\ b\sum c_{1i}W_{yi} \end{bmatrix}, \qquad (\boldsymbol{p}_{\text{equiv}})_2 = \begin{bmatrix} \sum l_i W_{xi}/L \\ a\sum c_{2i}W_{yi} \\ -b\sum c_{2i}W_{yi} \end{bmatrix}$$

with

$$c_{1i} = L(1-\epsilon) - l_i, \qquad c_{2i} = l_i - \epsilon L$$

the summations being taken over all the applied loads. These equations are easily obtained from statics. Details of their derivation have been given by the author elsewhere (Livesley, 1973). Transformation of equations (7.1) into global coordinates follows the usual pattern.

The assumptions of simple rigid-plastic behaviour imply that the member shown in Fig. 7.3 remains undeformed until, at some point, the moment or the axial force reaches its limiting value. This implies that there are "direct" constraints on the individual variables t, $\tilde{m}_1$ and $\tilde{m}_2$, corresponding to axial yield or the appearance of plastic hinges at the points A or B. These constraints may be written $-\boldsymbol{r}^L \leqslant \boldsymbol{r} \leqslant \boldsymbol{r}^U$. We shall assume that the components or $\boldsymbol{r}^L$ and $\boldsymbol{r}^U$ are all non-negative.

As well as these "direct" constraints there may be "indirect" constraints imposed on $\boldsymbol{r}$ because of moments at other points in the member. For example, the bending moment m_i at the point I in Fig. 7.3 is a linear function of $\boldsymbol{r}$ and the applied loads, and is limited by the value of the fully plastic moment at I. If there are a number of such critical points along the length of a member we shall have a series of moments . . . m_i . . ., each one being a linear function of $\boldsymbol{r}$ and the applied loads, and each one being limited by a constraint of the form $-m_i^L \leqslant m_i \leqslant m_i^U$.

From statics we have

$$\begin{aligned} m_i = \{ac_{2i} \sum_j c_{1j} W_{yj} - \sum_{i>j} (l_i - l_j) W_{yj}\} \\ -[0 \quad -ac_{1i} \quad -ac_{2i}]\boldsymbol{r} \end{aligned} \tag{7.2}$$

where $\sum_j$ indicates summation over all the loads on the member and $\sum_{i>j}$ indicates summation over those loads for which $l_i > l_j$. This may be written as

$$m_i = m_i^{(0)} - \boldsymbol{b}_i^t \boldsymbol{r}. \tag{7.3}$$

The term $m_i^{(0)}$ represents the moment which would occur at I if the beam were simply supported at A and B; this is often referred to as the "free" bending moment.†

The procedure described in Section 6.1 may be used to assemble the joint equilibrium equations for the complete structure

$$\mathbf{p} = \mathbf{Hr} \tag{7.4}$$

from the member equilibrium equations (7.1). The auxiliary equilibrium equations (7.3) for loads applied at points other than joints may also be collected together and written in the form

$$\mathbf{m} = \mathbf{m}^{(0)} - \mathbf{Br}. \tag{7.5}$$

(Note that equations (7.5) are not altered by a change to global coordinates.)

For a rigid-jointed plane frame with n joints, m members and s loads at points other than joints, $\mathbf{p}$ has $3n$ components, $\mathbf{r}$ has $3m$ components and $\mathbf{m}^{(0)}$ and $\mathbf{m}$ each have s components. Equations (7.4) and (7.5) may be combined and written as

† Strictly $m_i^{(0)}$ represents the moment at point I when $\tilde{m}_1$ and $\tilde{m}_2$ are zero.

$$\begin{matrix}3n\\s\end{matrix}\begin{bmatrix}\mathbf{p}\\\mathbf{m}^{(0)}\end{bmatrix}=\underset{\displaystyle 3m\qquad s}{\begin{bmatrix}\mathbf{H}&\mathbf{0}\\\mathbf{B}&\mathbf{I}\end{bmatrix}}\begin{bmatrix}\mathbf{r}\\\mathbf{m}\end{bmatrix}\tag{7.6a}$$

The constraints on the various components of **r** and **m** may be written in the form

$$\begin{bmatrix}-\mathbf{r}^L\\-\mathbf{m}^L\end{bmatrix}\leqslant\begin{bmatrix}\mathbf{r}\\\mathbf{m}\end{bmatrix}\leqslant\begin{bmatrix}\mathbf{r}^U\\\mathbf{m}^U\end{bmatrix}\tag{7.7a}$$

(Note that the form of the constraints given in (7.7a) is actually more general than that specified in Section 7.1 since it allows for fully plastic moments which have different values in the positive and negative senses.)

As an example, consider the frame shown in Fig. 7.4. Equation (7.6a) takes the form

$$\begin{matrix}13\\\text{scalar}\\\text{equa-}\\\text{tions}\end{matrix}\begin{bmatrix}p_A\\p_B\\p_C\\m_D^{(0)}\\m_E^{(0)}\\m_F^{(0)}\\m_G^{(0)}\end{bmatrix}=\underset{\text{19 scalar columns}}{\left[\begin{array}{ccccc|cccc}H_{2a}'&H_{1b}'&0&0&0&&&&\\0&H_{2b}'&H_{2c}'&H_{1d}'&0&&\mathbf{0}&&\\0&0&0&H_{2d}'&H_{1f}'&&&&\\\hline 0&b_D^t&0&0&0&1&0&0&0\\0&b_E^t&0&0&0&0&1&0&0\\0&0&0&b_F^t&0&0&0&1&0\\0&0&0&b_G^t&0&0&0&0&1\end{array}\right]}\begin{bmatrix}r_a\\r_b\\r_c\\r_d\\r_f\\m_D\\m_E\\m_F\\m_G\end{bmatrix}$$

So far we have said little about deformations. For each member stress-resultant vector $\boldsymbol{r}$ there will be a deformation vector $\boldsymbol{e}$ whose components correspond to $\boldsymbol{r}$ in a work sense. These components are the axial extension

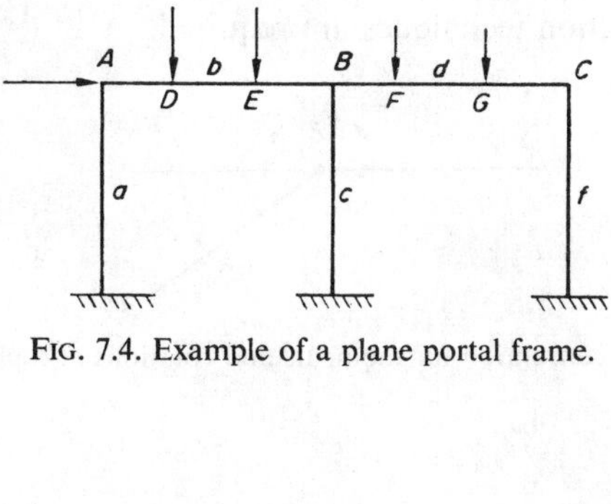

FIG. 7.4. Example of a plane portal frame.

e and the hinge rotations $\tilde{\theta}_1$, $\tilde{\theta}_2$, where the sense of positive rotation is defined by Fig. 7.5. The individual deformation vectors may be combined into a vector **e** in the usual way. For each moment m_i there will be a corresponding rotation θ_i, and these rotations may be combined to form a vector $\boldsymbol{\theta}$. Applying the usual virtual work argument to (7.6a) we find that the vectors **e** and $\boldsymbol{\theta}$ are related to the joint displacements **d** by the compatibility equations

$$\begin{matrix} 3m \\ s \end{matrix} \quad \begin{bmatrix} \mathbf{e} \\ \boldsymbol{\theta} \end{bmatrix} = \underset{\begin{matrix} 3n & \quad s \end{matrix}}{\begin{bmatrix} \mathbf{H}^t & \mathbf{B}^t \\ \mathbf{0} & \mathbf{I} \end{bmatrix}} \begin{bmatrix} \mathbf{d} \\ \boldsymbol{\theta} \end{bmatrix} \tag{7.8a}$$

Having established the technique for setting up the complete equations (7.6a) and (7.8a) it is convenient to absorb $\mathbf{m}^{(0)}$ in **p** and **m** in **r**. Similarly we absorb $\boldsymbol{\theta}$ in both **e** and **d**, and redefine **H** as the *complete* matrix appearing in (7.6a). If we do this equations (7.6a) and (7.8a) become the familiar equations

$$\mathbf{p} = \mathbf{Hr}, \quad \mathbf{e} = \mathbf{H}^t\mathbf{d} \qquad \text{(7.6b), (7.8b)}$$

while the constraints (7.7a) become simply

$$-\mathbf{r}^L \leqslant \mathbf{r} \leqslant \mathbf{r}^U \tag{7.7b}$$

It is apparent that the approach developed above may also be applied to continuum analysis. The Tresca yield criterion gives linear constraints on the principal stress differences. If the principal stresses are in known constant directions this leads to linear constraints on the element stress-resultant vectors, so that the techniques described later in this chapter may be used. If the directions of the principal stresses are unknown or the von Mises criterion is applied the constraints become non-linear, and more complex solution techniques are required.

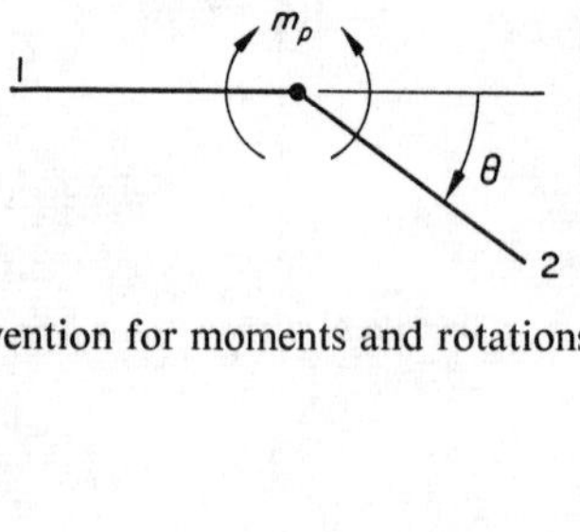

FIG. 7.5. Sign convention for moments and rotations at a plastic hinge.

7.3. Plastic collapse under proportional loading

We now consider the first of the problems listed at the beginning of this chapter. Given a structure under known applied loads, what is the factor λ_c by which the loads should be multiplied to bring the structure to the point of collapse? This value of λ is known as the *collapse load factor*. In the case of a determinate structure the problem is trivial, since the equilibrium equations (7.6a) may be solved explicitly for $\mathbf{r}$ and $\mathbf{m}$. Multiplying the loads by λ multiplies each component of $\mathbf{r}$ and $\mathbf{m}$ by λ, and the collapse load is simply the value of λ corresponding to the first appearance of yield.

In a hyperstatic structure the collapse load factor λ_c is likely to be considerably higher than the load factor at which yield first appears. To find λ_c we invoke the *lower bound theorem* of plasticity. This states that if in a given structure a system of internal forces and moments can be found which are in equilibrium with the applied loads and do not violate the yield constraints then the structure will support the loads. The theorem may be stated in the notation of equations (7.6b) and (7.7b) as

> "If a vector of internal stress resultants $\mathbf{r}$ can be found satisfying the yield constraints $-\mathbf{r}^L \leqslant \mathbf{r} \leqslant \mathbf{r}^U$ and the equilibrium equations $\lambda\mathbf{p} = \mathbf{H}\mathbf{r}$ then $\lambda \leqslant \lambda_c$."

It follows that the collapse load factor λ_c is the maximum value of λ for which a solution of the equilibrium equations exists which satisfies the yield constraints. The problem is therefore to maximize λ subject to $\lambda\mathbf{p} = \mathbf{H}\mathbf{r}$ and $-\mathbf{r}^L \leqslant \mathbf{r} \leqslant \mathbf{r}^U$. This is the "bounded variables problem" which was first defined and solved by Charnes and Lemke in 1954.

A direct method of solving this problem is given in Section 7.7. Alternatively the problem may be converted to standard linear programming form as follows. We first write it as

$$\left.\begin{aligned} &\text{“Maximize} \quad && w = [1 \quad \mathbf{0}] \begin{bmatrix} \lambda \\ \mathbf{r} \end{bmatrix} \\ &\text{subject to} && \begin{bmatrix} \mathbf{p} & -\mathbf{H} \\ \mathbf{0} & \mathbf{I} \\ \mathbf{0} & -\mathbf{I} \end{bmatrix} \begin{bmatrix} \lambda \\ \mathbf{r} \end{bmatrix} \begin{matrix} = \\ \leqslant \\ \leqslant \end{matrix} \begin{bmatrix} \mathbf{0} \\ \mathbf{r}^U \\ \mathbf{r}^L \end{bmatrix} \text{”} \end{aligned}\right\} \quad (7.9a)$$

We change to variables which are all positive by writing each r in the form $r = r^+ - r^-$, where $r = r^+$, $r^- = 0$ if $r \geqslant 0$ and $r = -r^-$, $r^+ = 0$ if $r \leqslant 0$. We combine all these equations, writing them as $\mathbf{r} = \mathbf{r}^+ - \mathbf{r}^-$, with the result that the problem becomes

$$\left.\begin{aligned} &\text{“Maximize} \quad w = [1 \;\; \mathbf{0} \;\; \mathbf{0}] \begin{bmatrix} \lambda \\ \mathbf{r}^+ \\ \mathbf{r}^- \end{bmatrix} \\ &\text{subject to} \quad \begin{bmatrix} \mathbf{p} & -\mathbf{H} & \mathbf{H} \\ \mathbf{0} & \mathbf{I} & -\mathbf{I} \\ \mathbf{0} & -\mathbf{I} & \mathbf{I} \end{bmatrix} \begin{bmatrix} \lambda \\ \mathbf{r}^+ \\ \mathbf{r}^- \end{bmatrix} \begin{matrix} = \\ \leqslant \\ \leqslant \end{matrix} \begin{bmatrix} \mathbf{0} \\ \mathbf{r}^U \\ \mathbf{r}^L \end{bmatrix} \text{”} \end{aligned}\right\} \tag{7.9b}$$

This is now in standard form. It will be noticed that the transformation of the problem greatly increases the size of matrix which has to be handled. If we consider the example shown in Fig. 7.4, the matrix **H** has 13 rows and 19 columns, while the matrix in (7.9b) has 51 rows and 39 columns. It would be unfair to suggest that this necessarily represents a proportionate increase in computer storage or processing time, since many of the coefficients are either 0 or 1. However, it does point to the need for efficient handling of zeros in any linear-programming computer package used for structural problems.

7.4. Duality and the mechanism approach

The reader with experience in the manual calculation of plastic collapse loads may wonder why we have chosen a lower bound approach, that is, an approach based on equilibrium considerations. We now show that the more familiar mechanism approach leads to a linear-programming problem which is the dual of the one formulated in the previous section.

In the mechanism approach we use the *upper bound theorem* of plasticity. This states that if in a given structure a compatible system of deformations can be found for which the work done by the applied loads is greater than the work done in plastic deformation then the structure will not support the loads.

We consider a set of displacements **d** and deformations **e** which are compatible, i.e. satisfy the equation $\mathbf{e} = \mathbf{H}^t\mathbf{d}$. To obtain an expression for the work done in plastic deformation we write each deformation component e in the form $e = e^+ - e^-$, where $e = e^+$, $e^- = 0$ if $e \geqslant 0$ and $e = -e^-$, $e^+ = 0$ if $e \leqslant 0$. We combine all these equations, writing them in the form $\mathbf{e} = \mathbf{e}^+ - \mathbf{e}^-$. Since $r = r^U$ if $e > 0$ and $r = -r^L$ if $e < 0$, the work done in plastic deformation can be written as $(\mathbf{r}^U)^t\mathbf{e}^+ + (\mathbf{r}^L)^t\mathbf{e}^-$. The work done by the applied loads is $\lambda\mathbf{p}^t\mathbf{d}$. Hence the upper bound theorem can be stated as

"If $\lambda\mathbf{p}^t\mathbf{d} \geqslant (\mathbf{r}^U)^t\mathbf{e}^+ + (\mathbf{r}^L)\mathbf{e}^-$, where $\mathbf{d}$, $\mathbf{e}^+$ and $\mathbf{e}^-$ satisfy the compatibility equations $\mathbf{e} = \mathbf{e}^+ - \mathbf{e}^- = \mathbf{H}^t\mathbf{d}$, then $\lambda \geqslant \lambda_c$."

The determination of λ_c may be put in standard linear programming form as follows. Since the deformation may be of arbitrary magnitude we may make the work input $\lambda\mathbf{p}^t\mathbf{d}$ equal to λ, i.e. we may impose the condition $\mathbf{p}^t\mathbf{d} = 1$. We may also write the equality $\mathbf{e}^+ - \mathbf{e}^- = \mathbf{H}^t\mathbf{d}$ as a pair of inequalities. Thus λ_c is the solution of the problem

"Minimize λ subject to $\lambda \geqslant (\mathbf{r}^U)^t\ \mathbf{e}^+ + (\mathbf{r}^L)^t\ \mathbf{e}^-$ and

$$\begin{bmatrix} \mathbf{p}^t & \mathbf{0} & \mathbf{0} \\ -\mathbf{H}^t & \mathbf{I} & -\mathbf{I} \\ \mathbf{H}^t & -\mathbf{I} & \mathbf{I} \end{bmatrix} \begin{bmatrix} \mathbf{d} \\ \mathbf{e}^+ \\ \mathbf{e}^- \end{bmatrix} \begin{matrix} = \\ \geqslant \\ \geqslant \end{matrix} \begin{bmatrix} 1 \\ \mathbf{0} \\ \mathbf{0} \end{bmatrix}\text{''}$$

Since the minimum value of λ obviously occurs when the first inequality is satisfied as an equality this may be written as

$$\left.\begin{aligned} &\text{“Minimize } \bar{w} = \lambda = [\mathbf{0}\ \ (\mathbf{r}^U)^t\ \ (\mathbf{r}^L)^t] \begin{bmatrix} \mathbf{d} \\ \mathbf{e}^+ \\ \mathbf{e}^- \end{bmatrix} \\ &\text{subject to } \begin{bmatrix} \mathbf{p}^t & \mathbf{0} & \mathbf{0} \\ -\mathbf{H}^t & \mathbf{I} & -\mathbf{I} \\ \mathbf{H}^t & -\mathbf{I} & \mathbf{I} \end{bmatrix} \begin{bmatrix} \mathbf{d} \\ \mathbf{e}^+ \\ \mathbf{e}^- \end{bmatrix} \begin{matrix} = \\ \geqslant \\ \geqslant \end{matrix} \begin{bmatrix} 1 \\ \mathbf{0} \\ \mathbf{0} \end{bmatrix}\text{''} \end{aligned}\right\} \tag{7.10}$$

The two problems posed in (7.9b) and (7.10) are dual problems, and it follows that Maximum w = Minimum $\bar{w}$. Thus the two approaches give exactly the same value of the collapse load-factor, as we should expect. At first sight it seems as though the equilibrium formulation (7.9b) gives the distribution of internal forces and moments at collapse but not the collapse mechanism. However, solution of (7.9b) by the standard Simplex procedure automatically produces as a by-product the values of the variables in the dual problem—that is, the values of $\mathbf{d}$, $\mathbf{e}^+$ and $\mathbf{e}^-$ which define the collapse mechanism. In a similar way the distribution of internal forces and moments at collapse may be obtained from the solution of (7.10).

In other sections of this book we have regarded $\mathbf{H}$ as being defined, in the first place, by equilibrium conditions rather than compatibility conditions. However, it is just as easy to set up $\mathbf{H}^t$ as $\mathbf{H}$, and we cannot really claim that (7.9b) is in any fundamental sense a "better" formulation than (7.10). From a computational point of view the choice between (7.9b) and (7.10) is likely to depend on the type of algorithm used to solve the linear programming problem.

7.5. The shakedown problem

Many structures are subjected to varying loads, and these can produce either *shakedown* or *incremental collapse*. These phenomena occur when a number of different loadings applied sequentially each produce a certain amount of yielding, although no one loading is sufficient to generate a full collapse mechanism. If the yielding continues indefinitely with each loading cycle, incremental collapse is said to occur. In effect, a collapse mechanism forms, although not all the plastic deformation takes place at the same time. If yielding eventually ceases, so that the ultimate behaviour is entirely elastic, the structure is said to have shaken down.

The shakedown load-factor λ_s is defined in a similar manner to the plastic collapse load-factor λ_c. If a sequence of loads $\mathbf{p}_1, \mathbf{p}_2, \ldots$ acts on a structure then the factor λ_s is defined as the factor by which these loads must be multiplied to just produce incremental collapse. The determination of λ_s is based on the *shakedown theorem*, which in effect says that if a state of a structure exists in which it can respond purely elastically to all loadings, it will find that state. More precisely, it states that a structure will shake down under load variations if a set of internal self-equilibrating stresses can be found such that any of the applied loads can be carried by purely elastic behaviour.

From equation (7.6b) it is clear that any set of internal stress-resultants $\mathbf{r}$ which satisfies

$$\mathbf{Hr} = \mathbf{0} \tag{7.11}$$

is self-equilibrating—that is, it can exist in the structure without any external loads being applied. If we carry out elastic analyses of the structure under all the sets of applied loads we can compute two vectors, consisting of the extreme values of the individual stress-resultants. Let these vectors be $\mathbf{r}^{\max}$ and $\mathbf{r}^{\min}$. The elastic calculations assume an initial unstressed state: if the initial state is defined by a vector $\mathbf{r}$ then the corresponding maximum and minimum elastic values will be $\mathbf{r}^{\max} + \mathbf{r}$ and $\mathbf{r}^{\min} + \mathbf{r}$. The theorem quoted above states that the structure will shake down if we can find an $\mathbf{r}$ satisfying (7.11) such that

$$-\mathbf{r}^L \leqslant \mathbf{r}^{\min} + \mathbf{r}, \quad \mathbf{r}^{\max} + \mathbf{r} \leqslant \mathbf{r}^U.$$

If we multiply each set of applied loads by λ, we also multiply $\mathbf{r}^{\max}$ and $\mathbf{r}^{\min}$ by λ, so that the calculation of the incremental collapse load-factor leads to the problem,

"Maximize λ subject to

$$\mathbf{Hr} = \mathbf{0}, \quad -\mathbf{r}^L \leqslant \lambda\mathbf{r}^{\min} + \mathbf{r}, \quad \lambda\mathbf{r}^{\max} + \mathbf{r} \leqslant \mathbf{r}^U."$$

This may be put in standard linear programming form as

$$\text{"Maximize} \quad w = [1 \quad \mathbf{0}] \begin{bmatrix} \lambda \\ \mathbf{r} \end{bmatrix}$$

$$\text{subject to} \quad \begin{bmatrix} 0 & \mathbf{H} \\ \mathbf{r}^{\max} & \mathbf{I} \\ -\mathbf{r}^{\min} & -\mathbf{I} \end{bmatrix} \begin{bmatrix} \lambda \\ \mathbf{r} \end{bmatrix} \begin{matrix} = \\ \leqslant \\ \leqslant \end{matrix} \begin{bmatrix} \mathbf{0} \\ \mathbf{r}^U \\ \mathbf{r}^L \end{bmatrix}\text{"}, \tag{7.12}$$

which is very similar in form to (7.9a). As before, $\mathbf{r}$ may be written as $\mathbf{r}^+ - \mathbf{r}^-$ to make all variables positive.

7.6. Minimum-weight design for single and multiple loadings

The general problem of optimum structural design is a complex one and has been the subject of much research work in recent years (see, for example, Fox (1971) and Gallagher and Zienkiewicz (1973)). In this section we simply demonstrate how certain problems in plastic design may be set up as linear programming problems. This approach involves a number of linearizing assumptions, which must be borne in mind when assessing numerical solutions. However, a linear programming problem is so much easier to solve than a non-linear one that an approximate optimum based on linearized theory may be "better", in an overall-cost sense, than a more exact solution.

The problem we consider is that of choosing member cross-sections for a plane frame or truss of given layout, in such a way that it will have a given collapse load-factor under some fixed or variable loading. In general there are likely to be many designs which satisfy this criterion. To produce a problem with a unique solution we impose the condition that the design must be, in some sense, the "best" one. The practical engineer's criterion for selecting the best design is likely to be that of cost. However, real costs are usually non-linear and discontinuous functions of the design variables. We therefore choose the weight of a structure as the function which defines its merit. If we consider a frame or truss with straight members of constant cross-section which are all made of the same material this function is $\rho A_i L_i$, the summation being over all the members.

In practice a designer may decide that several members must have the same cross-section, so that the number of independent design parameters is likely to be less than the total number of members. Let the set of independent cross-sectional areas be $A_1, A_2, \ldots$ and let $a_j = \rho A_j$. Then the weight is $W = a_j L_j$, where the value of L_j is the sum of the lengths of the members whose cross-section is defined by the corresponding a_j. We assume that the a_j's are continuous variables.

We now make the assumption that the limits t^L, t^U, m_p for each member are linear functions of the appropriate parameter a. This is reasonable in the case of the tensile force limit t^U, but is clearly an approximation in the case of t^L, since this limit is likely to be imposed by buckling considerations. The assumption that m_p is a linear function of cross-sectional area is also an approximation—for sections of similar shape m_p is actually proportional to $A^{3/2}$. These assumptions allow us to write the limits $\mathbf{r}^L$ and $\mathbf{r}^U$ as

$$\mathbf{r}^L = \mathbf{S}^L \boldsymbol{a}, \quad \mathbf{r}^U = \mathbf{S}^U \boldsymbol{a}.$$

The matrices $\mathbf{S}^L$, $\mathbf{S}^U$ will usually have a very simple form. For example, in the design of the two-bay portal shown in Fig. 7.4 we may decide that for fabrication reasons both beams are to have the same cross-sectional area A_b and that all the columns are to have the same cross-sectional area A_c. This gives

$$\boldsymbol{a} = \rho \begin{bmatrix} A_b \\ A_c \end{bmatrix},$$

so that $\mathbf{S}^L$ and $\mathbf{S}^U$ are each matrices with 19 rows and 2 columns. Each row contains a single coefficient, which associates the corresponding component of $\mathbf{r}$ with either α_b or α_c.

Having made these assumptions, our problem is to minimize $W = \mathbf{L}^t\boldsymbol{a}$ subject to the equilibrium equations $\mathbf{p} = \mathbf{Hr}$ and the yield constraints $\mathbf{S}^L\boldsymbol{a} \leqslant \mathbf{r} \leqslant \mathbf{S}^U\boldsymbol{a}$. This may be written in standard linear programming form as

$$\text{“Minimize } W = [\mathbf{0} \quad \mathbf{L}^t] \begin{bmatrix} \mathbf{r} \\ \boldsymbol{a} \end{bmatrix}$$

$$\text{subject to} \quad \begin{bmatrix} \mathbf{H} & \mathbf{0} \\ \mathbf{I} & \mathbf{S}^L \\ -\mathbf{I} & \mathbf{S}^U \end{bmatrix} \begin{bmatrix} \mathbf{r} \\ \boldsymbol{a} \end{bmatrix} \begin{matrix} = \\ \geqslant \\ \geqslant \end{matrix} \begin{bmatrix} \mathbf{p} \\ \mathbf{0} \\ \mathbf{0} \end{bmatrix}\text{”} \tag{7.13}$$

As before, $\mathbf{r}$ may be replaced by $\mathbf{r}^+ - \mathbf{r}^-$ to make all variables positive (the components of $\boldsymbol{a}$ are essentially positive anyway).

We now consider the case where $\mathbf{p}$ is not a fixed loading but may vary between certain limits. Keeping within the restrictions imposed by linearity, we assume that $\mathbf{p}$ is a linear combination of certain basic loading systems $\mathbf{p}_1, \mathbf{p}_2, \ldots$, so that $\mathbf{p} = \beta_i\mathbf{p}_i$. Combining the various loadings $\mathbf{p}_i$ into a matrix $\mathbf{P}$ gives $\mathbf{p} = \mathbf{P}\boldsymbol{\beta}$. If we imagine that the multipliers β_i are restricted by some system of linear constraints $\mathbf{C}\boldsymbol{\beta} \leqslant \mathbf{f}$, our problem is

$$\text{“Minimize } W = [\mathbf{0} \quad \mathbf{L}^t \quad \mathbf{0}] \begin{bmatrix} \mathbf{r} \\ \boldsymbol{a} \\ \boldsymbol{\beta} \end{bmatrix}$$

$$\text{subject to} \quad \begin{bmatrix} \mathbf{H} & \mathbf{0} & -\mathbf{P} \\ \mathbf{I} & \mathbf{S}^L & \mathbf{0} \\ -\mathbf{I} & \mathbf{S}^U & \mathbf{0} \\ \mathbf{0} & \mathbf{0} & \mathbf{C} \end{bmatrix} \begin{bmatrix} \mathbf{r} \\ \boldsymbol{a} \\ \boldsymbol{\beta} \end{bmatrix} \begin{matrix} = \\ \geqslant \\ \geqslant \\ \leqslant \end{matrix} \begin{bmatrix} \mathbf{0} \\ \mathbf{0} \\ \mathbf{0} \\ \mathbf{f} \end{bmatrix}\text{”} \tag{7.14}$$

Once again, we see that a formal linear-programming formulation gives us a large sparse matrix.

This approach produces a design that will not collapse under any *single* loading within the prescribed limits. It does not, however, ensure that the design will shake down under any *sequence* of loads within the limits. Design (as opposed to analysis) for shakedown is essentially a non-linear process and, as such, outside the scope of this chapter. It has been considered by Heyman (1958).

In addition to the linearizing assumptions mentioned earlier we have assumed that all the design parameters α_j can be varied continuously. This clearly limits the usefulness of the approach as far as steel structures are concerned, since steel sections only come in standard sizes. This defect can be overcome by a technique known as integer programming, but the complexity of this technique makes it of doubtful practical value.

7.7. A compact procedure for finding the collapse load factor

In Section 7.3 we showed how the "bounded variables problem" can be converted to standard linear programming form. We also saw that this transformation leads to a considerable increase in the size of the coefficient matrix. We now describe a direct method of solving the original problem which is considerably more economical of computer storage. The reader familiar with numerical methods for solving the linear programming problem will recognize this method as being based on the Simplex procedure.

The problem is defined as, "Maximize λ subject to $\lambda\mathbf{p} = \mathbf{H}\mathbf{r}$ and $-\mathbf{r}^L \leqslant \mathbf{r} \leqslant \mathbf{r}^U$, where $\mathbf{p}$, $\mathbf{H}$, $\mathbf{r}^L$ and $\mathbf{r}^U$ are given". Let $\mathbf{p}$ and $\mathbf{r}$ have M and N components respectively, where $M < N$. (This implies that $\mathbf{H}$ is a matrix with M rows and N columns.)

We begin by applying the Gauss–Jordan reduction process described in Section 6 of the Appendix to the equilibrium equations $\lambda\mathbf{p} = \mathbf{H}\mathbf{r}$, taking the rows of $\mathbf{H}$ in turn and selecting the largest element in each row as pivotal element. This converts the equations to a set of equations $\lambda\mathbf{p}^* = \mathbf{H}^*\mathbf{r}$, which have the form,

$$\lambda\begin{bmatrix} p_1^* \\ \cdot \\ \cdot \\ \cdot \\ \cdot \\ \cdot \\ p_M^* \end{bmatrix} = \begin{bmatrix} 0 & . & 1 & . & . & 0 & . & & & & 0 \\ 1 & . & 0 & . & . & 0 & . & & & & 0 \\ 0 & . & 0 & . & . & 0 & . & & & & 1 \\ . & . & . & . & . & . & . & . & . & . & . \\ . & . & . & . & . & . & . & . & . & . & . \\ 0 & . & 0 & . & . & 1 & . & & & & 0 \\ 0 & . & 0 & . & . & 0 & . & & & & 0 \end{bmatrix} \begin{bmatrix} r_1 \\ \cdot \\ \cdot \\ \cdot \\ \cdot \\ \cdot \\ \cdot \\ \cdot \\ r_N \end{bmatrix} \tag{7.15}$$

with M of the N columns of $\mathbf{H}$ having been reduced to columns containing only a single 1. (This part of the process is also used in the method of automatic redundant selection described in Section 8.5).

We now divide the components of $\mathbf{r}$ into two types which (using the terminology of linear programming) we call *basic* and *non-basic* variables. For our purposes, a basic variable is one that satisfies the constraints $-r^L < r < r^U$ as a pair of strict inequalities, so that it can be altered by a finite amount without violating its constraints. Basic variables are associated with columns of $\mathbf{H}^*$ that have been reduced (i.e. contain only a single 1). A non-basic variable is one which is at one or other of its limits, except for an initial stage, when its value is zero. Non-basic variables are associated with columns of $\mathbf{H}^*$ which have no specific arrangement. By reducing the equilibrium equations to the form (7.15) we have effectively chosen M of the r_j's to be basic and $N - M$ to be non-basic. Note that each basic variable only appears in one equation, and each equation only contains one basic variable. Initially we put $\lambda = 0$ and all $r_j = 0$.

The remainder of the computational process consists of two alternating steps,

1. Increasing λ, altering only the basic variables, until one of them reaches one of its limiting values.
2. Interchanging a basic and a non-basic variable so that a further increase in λ becomes possible.

When step (2) cannot be carried out, the maximum value of λ has been reached.

Step 1: increasing λ.

Imagine that at a particular stage we have a value of λ equal to $\lambda^{(0)}$, and M basic variables $r_{i'}$ with values $r_{i'}^{(0)}$ satisfying $-r_{i'}^{L} < r_{i'}^{(0)} < r_{i'}^{U}$. (The suffix i' denotes the basic variable associated with equation i.) Consider now the effect of an increase $\delta\lambda$ in the load factor. Since each basic variable is strictly within its limits, equalities (7.15) may be satisfied by keeping all the non-basic variables constant and letting the basic variables vary. For equation i we have

$$p_i^*\delta\lambda = \delta r_{i'},$$

by virtue of the form of $\mathbf{H}^*$. Thus if p_i^* is positive $r_{i'}$ increases, while if p_i^* is negative $r_{i'}$ decreases. The limits on $r_{i'}$ mean that

$$\delta\lambda \leqslant (r_{i'}^{U} - r_{i'}^{(0)})/p_i^* \qquad \text{if } p_i^* > 0,$$

$$\delta\lambda \leqslant (-r_{i'}^{L} - r_{i'}^{(0)})/p_i^* \quad \text{if } p_i^* < 0.$$

If $p_i^* = 0$ then the value of $r_{i'}$ is not altered by a change in λ.

The smallest value of $\delta\lambda$ taken over all the equations determines the permissible increase in λ. Let the critical equation be k and the permissible increase $\delta\lambda_k$. Then we have

$$\lambda^{(1)} = \lambda^{(0)} + \delta\lambda_k, \quad r_{i'}^{(1)} = r_{i'}^{(0)} + p_i^*\delta\lambda_k \quad \text{(all basic variables)}$$

with $r_{k'}^{(1)} = -r_{k'}^{L}$ or $r_{k'}^{U}$ according to the sign of p_k^*, and all other basic variables still within their limits. The non-basic variables remain at their previous values.

Step 2: altering the set of basic variables.

It is apparent that to achieve any further increase in λ we must find some other variable to take over the role of basic variable for equation k. Equation k now has the form

$$\lambda^{(1)}p_k^* = h_{kj}^*r_j + r_{k'}$$

where j is a dummy suffix denoting all the non-basic variables. (All the basic variables, except for $r_{k'}$, have zero coefficients in this equation.) We need an r_j which can be varied in a permissible manner in such a way that λ will increase. Each non-basic r_j may be in one of three states,

1. $r_j^{(1)} = 0$ This is the initial state. Both positive and negative variations of r_j are (in general) permissible, and one of these will certainly increase λ, provided that $h_{kj}^* \neq 0$.

2. $r_j^{(1)} = -r_j^L$ Only an increase in r_j is permissible. This will increase λ provided that $h_{kj}^* \neq 0$ and is of the *same* sign as p_k^*.

3. $r_j^{(1)} = r_j^U$ Only a decrease in r_j is permissible. This will increase λ provided that $h_{kj}^* \neq 0$ and is of *different* sign to p_k^*.

It seems advantageous, if there is a choice, to choose the r_j which produces the largest rate of increase in λ. Hence we find the value of the index j which maximizes the quantity z_j, where

$$\begin{aligned} z_j &= |\, h_{kj}^* \,| & \text{if } r_j^{(1)} &= 0, \\ z_j &= h_{kj}^* \text{ sign } p_k^* & \text{if } r_j^{(1)} &= -r_j^L, \\ z_j &= -h_{kj}^* \text{ sign } p_k^* & \text{if } r_j^{(1)} &= r_j^U. \end{aligned}$$

If this index is l and if $z_l > 0$, then r_l is chosen as the new basic variable. We accordingly reduce column l to a column of zeros with a 1 in row k by carrying out a Gauss–Jordan step using h_{kl}^* as pivot. This eliminates the coefficients of r_l in all the other equations and generates a column of coefficients for $r_{k'}$. The equations are now in "standard form" again, so that we may return to consider the next increase in λ. In this increase the variable we have called $r_{k'}$, which is now non-basic, remains at the limit which it has just reached. If $z_l \leqslant 0$, no further increase in λ is possible and the solution has been reached. A simple extension of this algorithm produces the collapse mode as a by-product of the calculation (see Livesley, 1973).

It is straightforward to construct a computer program to assemble the matrix **H**, carry out the above calculation and plot the collapse mechanism and bending moment diagram. Experience with such a program suggests that the calculation of the collapse load-factor of a structure takes about 5 times as long as an elastic analysis of the same structure. While this ratio is obviously very much a function of the relative efficiency of specific computer programs, similar ratios have been found by others who have compared linear programming routines with routines for solving linear equations.

CHAPTER 8

The Compatibility or Force Method

We now return to the analysis of linear elastic structures. As we saw in Chapter 6, the equilibrium or displacement method of analysis requires the solution of the load/displacement equations

$$\mathbf{p} = \mathbf{H}\mathbf{K}_m\mathbf{H}^t\mathbf{d} \tag{8.1}$$

or

$$\mathbf{p} = \mathbf{K}\mathbf{d} \tag{8.2}$$

where the diagonal matrix $\mathbf{K}_m$ defines the stiffness of the individual disconnected elements and $\mathbf{H}$ is the matrix which relates the applied loads $\mathbf{p}$ to the element stress-resultants $\mathbf{r}$. Since the order of the matrix $\mathbf{K}$ is equal to the total number of nodal degrees of freedom of the structure the solution of equation (8.2) for a given load vector $\mathbf{p}$ (or the corresponding inversion process for a general load vector) may be regarded as the central computational step of the method.

The equilibrium approach has two features which in certain circumstances may be regarded as undesirable. In the first place the solution of (8.2) produces nodal displacements, which may be of less interest to the design engineer than internal stress-resultants. (The stress-resultants are, of course, obtainable from the displacements, but only at the cost of additional calculations.) In the second place the stiffness matrix $\mathbf{K}$ is not noticeably simpler in the case of a determinate structure, and indeed the procedure for assembling $\mathbf{K}$ described in Section 5.3 is independent of any considerations of determinacy.

We saw in Chapter 6 that in the case of a determinate structure the equilibrium matrix $\mathbf{H}$ is square and non-singular, so that the solution of

(8.1) may be split up into three steps:

$$\mathbf{r} = \mathbf{H}^{-1}\mathbf{p}, \tag{8.3a}$$

$$\mathbf{e} = \mathbf{F_m r}, \tag{8.3b}$$

$$\mathbf{d} = (\mathbf{H}^{-1})^t\mathbf{e}. \tag{8.3c}$$

In this case, of course, the element stress-resultants are independent of the element flexibilities and may be obtained immediately from (8.3a). The matrix **H** is usually of simple form, and we saw in Section 6.4 that for certain types of determinate structure it is possible to write down $\mathbf{H}^{-1}$ directly.

The relative simplicity of the analysis when applied to determinate structures suggests the possibility of finding a procedure which retains at least some part of this simplicity when applied to structures whose degree of indeterminacy is small. This is the idea behind the *compatibility* or *force* method.

In traditional presentations of the method the procedure is described as an imaginary physical process. A hyperstatic structure is first made determinate by "cutting" it at certain points. The loads are then applied to this determinate structure, which is analysed by means such as those described in Section 6.3. This analysis gives values for the "gaps" which the loading produces at the imaginary cuts. The final step is to find the "redundants", that is, the pairs of forces and moments which must be applied at the cuts to reduce all the gaps to zero.

This form of the method has an intuitive appeal and several examples of its use are presented in this chapter. As a basis for general computer programs, however, it introduces a new problem—that of finding rules for selecting the number and positions of the imaginary cuts. While it is possible to develop such rules it is more satisfactory to return to the original equilibrium equations and redevelop the analysis from a purely algebraic point of view. In Section 8.5 we describe an algorithm which automatically selects an appropriate set of "redundants" **q** and converts the equations

$$\mathbf{p} = \mathbf{Hr}$$

into the form

$$\mathbf{r} = \mathbf{B}_0\mathbf{p} + \mathbf{Bq}. \tag{8.4}$$

This "inverse" form of the equilibrium equations generates a corresponding set of compatibility equations. The remaining steps of the method may also be justified by a virtual work argument which does not involve consideration of imaginary "cuts" or "gaps" in the structure. Thus we obtain a statement of the compatibility method which (like the equilibrium method) makes no demands on human judgement.

8.1. The analysis of a pin-jointed truss

For our first illustration of the method we consider the plane pin-jointed truss shown in Fig. 8.1. This is the structure we considered in Section 6.1,

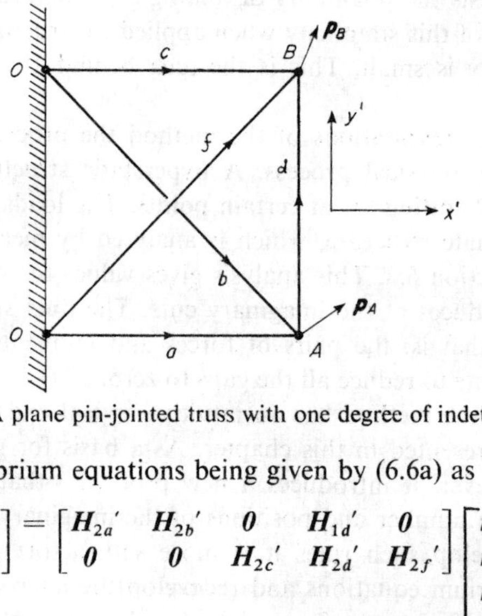

FIG. 8.1. A plane pin-jointed truss with one degree of indeterminacy.

the nodal equilibrium equations being given by (6.6a) as

$$\begin{bmatrix} \boldsymbol{p}_A \\ \boldsymbol{p}_B \end{bmatrix} = \begin{bmatrix} \boldsymbol{H}_{2a}' & \boldsymbol{H}_{2b}' & \boldsymbol{0} & \boldsymbol{H}_{1d}' & \boldsymbol{0} \\ \boldsymbol{0} & \boldsymbol{0} & \boldsymbol{H}_{2c}' & \boldsymbol{H}_{2d}' & \boldsymbol{H}_{2f}' \end{bmatrix} \begin{bmatrix} r_a \\ r_b \\ r_c \\ r_d \\ r_f \end{bmatrix} \tag{8.5}$$

For a pin-ended member in a plane structure

$$\boldsymbol{H}_1' = -\begin{bmatrix} \cos\alpha \\ \sin\alpha \end{bmatrix}, \qquad \boldsymbol{H}_2' = \begin{bmatrix} \cos\alpha \\ \sin\alpha \end{bmatrix},$$

so that the equilibrium equations (8.5) may be written out explicitly as

$$\begin{bmatrix} p_{xA} \\ p_{yA} \\ p_{xB} \\ p_{yB} \end{bmatrix} = \begin{bmatrix} 1 & 1/\sqrt{2} & 0 & 0 & 0 \\ 0 & -1/\sqrt{2} & 0 & -1 & 0 \\ 0 & 0 & 1 & 0 & 1/\sqrt{2} \\ 0 & 0 & 0 & 1 & 1/\sqrt{2} \end{bmatrix} \begin{bmatrix} r_a \\ r_b \\ r_c \\ r_d \\ r_f \end{bmatrix} \tag{8.6}$$

where $r_a \ldots r_f$ are the tensions in the members. These equations may easily be checked by statics. They are identical with (6.14) except for the extra column of **H** associated with member f and the extra component r_f in the vector of member tensions. The addition of the extra member means that we cannot solve (8.6), since there are now five unknowns and only four equations. (The difference $5 - 4 = 1$ gives us the "degree of indeterminacy" of the structure.)

The next step is to temporarily disconnect a member from one of the joints to which it is attached. In this example we disconnect end 2 of member f from joint B, this choice being arbitrary. We apply a pair of equal and opposite forces of magnitude q to the two sides of the gap, as shown in Fig. 8.2. This process is sometimes described as inserting a *release* in the structure. Note that the pair of forces associated with the release is essentially self-equilibrating.

We may now add the equation $q = -r_f$ to the equilibrium equations (8.6), obtaining

$$\begin{bmatrix} p_{xA} \\ p_{yA} \\ p_{xB} \\ p_{yB} \\ q \end{bmatrix} = \begin{bmatrix} 1 & 1/\sqrt{2} & 0 & 0 & 0 \\ 0 & -1/\sqrt{2} & 0 & -1 & 0 \\ 0 & 0 & 1 & 0 & 1/\sqrt{2} \\ 0 & 0 & 0 & 1 & 1/\sqrt{2} \\ 0 & 0 & 0 & 0 & -1 \end{bmatrix} \begin{bmatrix} r_a \\ r_b \\ r_c \\ r_d \\ r_f \end{bmatrix} \tag{8.7}$$

This has the effect of replacing **H** by an augmented matrix which we shall call $\mathbf{H}_+$, which is square and non-singular. Inverting (8.7) we obtain

$$\begin{bmatrix} r_a \\ r_b \\ r_c \\ r_d \\ r_f \end{bmatrix} = \left[\begin{array}{cccc|c} 1 & 1 & 0 & 1 & 1/\sqrt{2} \\ 0 & -\sqrt{2} & 0 & -\sqrt{2} & -1 \\ 0 & 0 & 1 & 0 & 1/\sqrt{2} \\ 0 & 0 & 0 & 1 & 1/\sqrt{2} \\ 0 & 0 & 0 & 0 & -1 \end{array}\right] \begin{bmatrix} p_{xA} \\ p_{yA} \\ p_{xB} \\ p_{yB} \\ q \end{bmatrix} \tag{8.8}$$

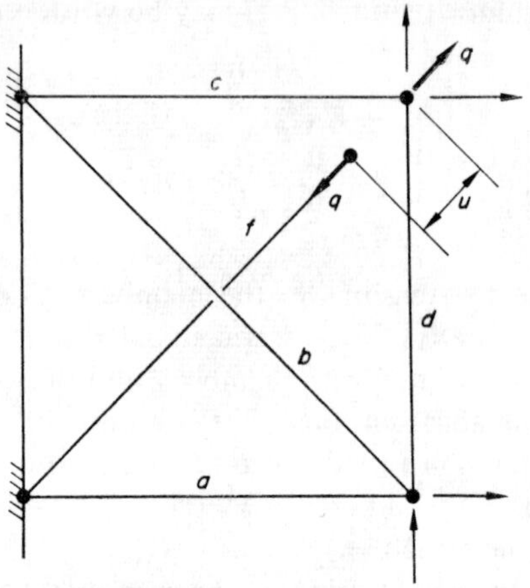

FIG. 8.2. The truss of Fig. 8.1 made statically determinate by the insertion of a release.

These equations are very similar to (6.15) and may easily be checked by statics. We split up the matrix $\mathbf{H}_{+}^{-1}$ appearing in (8.8) into two rectangular matrices, which we call $\mathbf{B}_0$ and $\mathbf{B}$. The matrix $\mathbf{B}_0$ consists of the first four columns of $\mathbf{H}_{+}^{-1}$, which are associated with the known applied loads $\boldsymbol{p}_A$ and $\boldsymbol{p}_B$. The matrix $\mathbf{B}$ consists of the last column, which is associated with the unknown load-pair q. Thus we can write (8.8) in the form

$$\mathbf{r} = \mathbf{B}_0\mathbf{p} + \mathbf{B}q. \tag{8.9}$$

Having expressed all the member tensions as linear functions of the known applied loads and the unknown q we can obtain the vector of member extensions $\mathbf{e}$ from the individual member flexibilities, writing

$$\mathbf{e} = \mathbf{F}_m\mathbf{r} = \mathbf{F}_m(\mathbf{B}_0\mathbf{p} + \mathbf{B}q) \tag{8.10}$$

where

$$\mathbf{F}_m = \begin{bmatrix} (L/EA)_a & 0 & 0 & 0 & 0 \\ 0 & (L/EA)_b & 0 & 0 & 0 \\ 0 & 0 & (L/EA)_c & 0 & 0 \\ 0 & 0 & 0 & (L/EA)_d & 0 \\ 0 & 0 & 0 & 0 & (L/EA)_f \end{bmatrix}$$

Having added the load-pair q to the vector of applied loads $\mathbf{p}$ it is necessary to add the corresponding displacement to the vector of joint displacements $\mathbf{d}$. The displacement which corresponds in a work sense to the pair of forces q, $-q$ is the *relative* displacement of their points of application, i.e. the displacement u in Fig. 8.2. Since (8.9) may be written as

$$\mathbf{r} = [\mathbf{B}_0 \quad \mathbf{B}] \begin{bmatrix} \mathbf{q} \\ q \end{bmatrix}$$

it follows by the usual virtual work argument that

$$\begin{bmatrix} \mathbf{d} \\ u \end{bmatrix} = \begin{bmatrix} \mathbf{B}_0{}^t \\ \mathbf{B}^t \end{bmatrix} \mathbf{e} \tag{8.11}$$

Combining (8.10) and (8.11) we obtain

$$\begin{bmatrix} \mathbf{d} \\ u \end{bmatrix} = \begin{bmatrix} \mathbf{B}_0{}^t \\ \mathbf{B}^t \end{bmatrix} \mathbf{F}_\mathrm{m}[\mathbf{B}_0\mathbf{p} + \mathbf{B}q] \tag{8.12}$$

If q were actually a known applied force this equation would simply be a special case of equation (6.12). However, q is an unknown which has to be determined from the condition $u = 0$. Expanding (8.12) we obtain

$$\left.\begin{aligned} \mathbf{d} &= \mathbf{B}_0{}^t\mathbf{F}_\mathrm{m}\mathbf{B}_0\mathbf{p} + \mathbf{B}_0{}^t\mathbf{F}_\mathrm{m}\mathbf{B}q, \\ u &= \mathbf{B}^t\mathbf{F}_\mathrm{m}\mathbf{B}_0\mathbf{p} \ + \ \mathbf{B}^t\mathbf{F}_\mathrm{m}\mathbf{B}q. \end{aligned}\right\} \tag{8.13}$$

Putting $u = 0$ gives

$$q = -(\mathbf{B}^t\mathbf{F}_\mathrm{m}\mathbf{B})^{-1}\mathbf{B}^t\mathbf{F}_\mathrm{m}\mathbf{B}_0\mathbf{p}. \tag{8.14}$$

Note that although written in matrix form the product $\mathbf{B}^t\mathbf{F}_\mathrm{m}\mathbf{B}$ is in this example a single number, so that its "inversion" is trivial.

Equation (8.14) may be given physical meaning as follows. We imagine first of all that the loads $\mathbf{p}$ are applied to the structure with member f not connected to B and carrying no load. Thus q is zero and the gap u is given by $u = \mathbf{B}^t\mathbf{F}_\mathrm{m}\mathbf{B}_0\mathbf{p}$. This is a straightforward application of equation (6.12), the matrices $\mathbf{B}_0$ and $\mathbf{B}$ being, as we have seen, merely the appropriate columns of the square matrix $(\mathbf{H}_+)^{-1}$. We now require the value of the pair of forces q, $-q$ which will make the gap zero. Once again we are dealing with a determinate structure and we find that the flexibility of this structure, as measured at the gap, is $\mathbf{B}^t\mathbf{F}_m\mathbf{B}$. Thus the required value of q is $-(\mathbf{B}^t\mathbf{F}_m\mathbf{B})^{-1}u$, which is equivalent to (8.14).

Having determined q we return to (8.9) to obtain the set of bar tensions

$$\mathbf{r} = (\mathbf{I} - \mathbf{B}(\mathbf{B}^t\mathbf{F}_m\mathbf{B})^{-1}\mathbf{B}^t\mathbf{F}_m)\mathbf{B}_0\mathbf{p}. \tag{8.15}$$

The displacements of the joints are given by

$$\mathbf{d} = \mathbf{B}_0{}^t\mathbf{e} = \mathbf{B}_0{}^t\mathbf{F}_m\mathbf{r} \tag{8.16}$$

while as a check on the analysis we may evaluate the product $\mathbf{B}^t\mathbf{F}_m\mathbf{r}$, which should be zero from (8.13).

The matrices $\mathbf{B}_0$ and $\mathbf{B}$ depend on the choice of q. In this example we could have placed the cut in any of the bars of the structure. Another choice would have given rise to different matrices, but for a given loading $\mathbf{p}$ the numerical values computed from (8.15) and (8.16) would have been the same.

8.2. The analysis of a rigid-jointed frame

For our second illustration of the method we consider the simple portal frame with rigid joints shown in Fig. 8.3. The loads at the joints A and B are assumed to include any equivalent loads arising from distributed loads on the members. As in Section 6.1, we shall work entirely in terms of the $\boldsymbol{F}$ and $\boldsymbol{H}'$ matrices of the members, so that provided these matrices and the vectors $\boldsymbol{p}$, $\boldsymbol{d}$, $\boldsymbol{r}$ and $\boldsymbol{e}$ are suitably defined our analysis will apply equally well to a plane or a space structure.

The equilibrium and compatibility methods have the same starting-point—the equations of joint equilibrium. Applying the rules given in Section 6.1 for the assembly of these equations we obtain

$$\begin{bmatrix} \boldsymbol{p}_A \\ \boldsymbol{p}_B \end{bmatrix} \doteq \begin{bmatrix} \boldsymbol{H}_{2a}{}' & \boldsymbol{H}_{1b}{}' & \boldsymbol{0} \\ \boldsymbol{0} & \boldsymbol{H}_{2b}{}' & \boldsymbol{H}_{1c}{}' \end{bmatrix} \begin{bmatrix} \boldsymbol{r}_b \\ \boldsymbol{r}_b \\ \boldsymbol{r}_c \end{bmatrix} \tag{8.17}$$

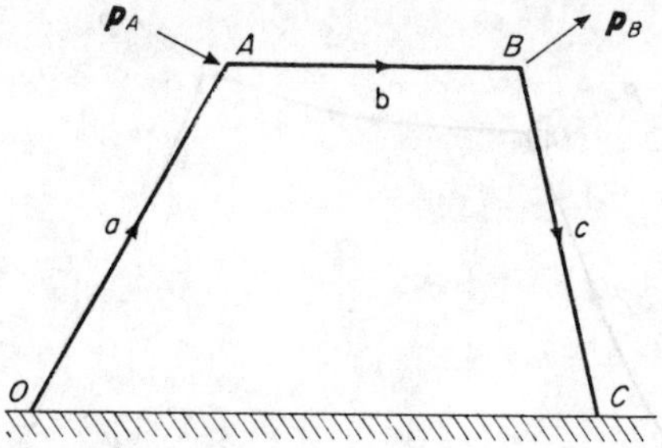

FIG. 8.3. A portal frame with three (or six) degrees of statical indeterminacy.

For comparison with the analysis presented in this section we note that the load/displacement equations set up in the equilibrium approach are

$$\begin{bmatrix} \boldsymbol{p}_A \\ \boldsymbol{p}_B \end{bmatrix} = \begin{bmatrix} (\boldsymbol{K}_{22}')_a + (\boldsymbol{K}_{11}')_b & (\boldsymbol{K}_{12}')_b \\ (\boldsymbol{K}_{21}')_b & (\boldsymbol{K}_{22}')_b + (\boldsymbol{K}_{11})'_c \end{bmatrix} \begin{bmatrix} \boldsymbol{d}_A \\ \boldsymbol{d}_B \end{bmatrix} \tag{8.18}$$

Equations (8.18) represents six scalar equations if the structure is a plane one and twelve scalar equations if it is three-dimensional.

As in the previous example, we see that the matrix **H** in (8.17) is rectangular. To make it square we consider the related structure shown in Fig. 8.4, in which the joint C is allowed to displace. As in the previous example, the pair of loads $\boldsymbol{q}$, $-\boldsymbol{q}$ applied to the two sides of the gap is self-equilibrating. The displacement $\boldsymbol{u}$ is again the *relative* displacement of the two sides of the gap, although in this case it happens to be equal to the absolute displacement of C. If the frame is a plane one the relaxation of the compatibility condition at C introduces three extra scalar load and displacement components into the vectors **p** and **d**, while if it is a space frame the number is six. We may therefore refer to the imaginary cut in the frame at C as a three-fold or six-fold release.

It is apparent from Fig. 8.4 that $\boldsymbol{q} = \boldsymbol{p}_{2c}' = \boldsymbol{H}_{2c}'\boldsymbol{r}_c$. Adding this equation to (8.17) we obtain

$$\begin{bmatrix} \boldsymbol{p}_A \\ \boldsymbol{p}_B \\ \boldsymbol{q} \end{bmatrix} = \begin{bmatrix} \boldsymbol{H}_{2a}' & \boldsymbol{H}_{1b}' & \boldsymbol{0} \\ \boldsymbol{0} & \boldsymbol{H}_{2b}' & \boldsymbol{H}_{1c}' \\ \boldsymbol{0} & \boldsymbol{0} & \boldsymbol{H}_{2c}' \end{bmatrix} \begin{bmatrix} \boldsymbol{r}_a \\ \boldsymbol{r}_b \\ \boldsymbol{r}_c \end{bmatrix} \tag{8.19}$$

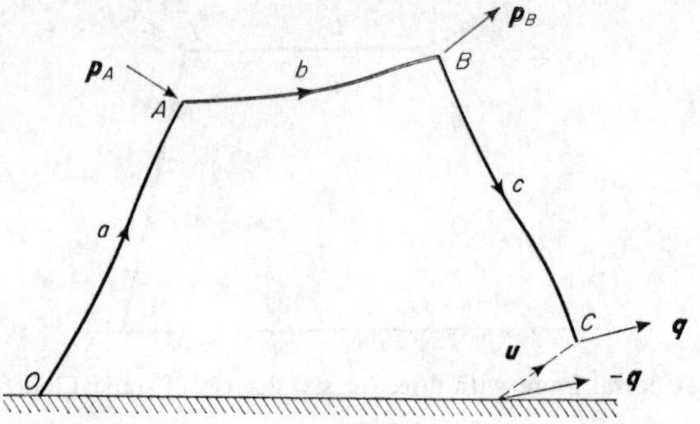

FIG. 8.4. The frame of Fig. 8.3 made statically determinate by the insertion of a release.

We may use the rule developed in Section 6.4 to obtain the inverse form of (8.19). The introduction of the cut at C has converted the frame into a tree structure with a single branch $OABC$. If all the member flexibility matrices and H' matrices are expressed in the coordinate system $Ox'y'$ the inverse equations are

$$\begin{bmatrix} r_a \\ r_b \\ r_c \end{bmatrix} = \left[\begin{array}{cc|c} (H_{2a}')^{-1} & (H_{2b}')^{-1} & (H_{2c}')^{-1} \\ 0 & (H_{2b}')^{-1} & (H_{2c}')^{-1} \\ 0 & 0 & (H_{2c}')^{-1} \end{array}\right] \begin{bmatrix} p_A \\ p_B \\ q \end{bmatrix} \tag{8.20}$$

As before, we split the matrix appearing in (8.20) into two matrices $\mathbf{B}_0$ and $\mathbf{B}$, the division being indicated by the dotted line. We write (8.20) as

$$\mathbf{r} = \mathbf{B}_0\mathbf{p} + \mathbf{B}\boldsymbol{q} \tag{8.21}$$

where $\mathbf{B}_0$ and $\mathbf{B}$ are 9×6 and 9×3 matrices respectively if the structure is being treated as a plane frame. (For a space frame the corresponding sizes are 18×12 and 18×6.)

The remainder of the analysis is almost identical with that given in the previous section. The compatibility equation corresponding to (8.21) is

$$\begin{bmatrix} \mathbf{d} \\ \boldsymbol{u} \end{bmatrix} = \begin{bmatrix} \mathbf{B}_0{}^t \\ \mathbf{B}^t \end{bmatrix} \mathbf{e} \tag{8.22}$$

while the equation connecting $\mathbf{e}$ and $\mathbf{r}$ is $\mathbf{e} = \mathbf{F_m r}$, where

$$\mathbf{F_m} = \begin{bmatrix} F_a & 0 & 0 \\ 0 & F_b & 0 \\ 0 & 0 & F_c \end{bmatrix}$$

Combining this equation with (8.21) and (8.22) gives

$$\mathbf{d} = \mathbf{B_0 F_m (B_0 p} + \mathbf{B}\boldsymbol{q}),$$

$$\boldsymbol{u} = \mathbf{BF_m(B_0 p} + \mathbf{B}\boldsymbol{q})$$

and applying the condition $\boldsymbol{u} = \boldsymbol{0}$ gives

$$\boldsymbol{q} = -(\mathbf{B}^t\mathbf{F_m B})^{-1}\mathbf{B}^t\mathbf{F_m B_0 p}.$$

Note that the matrix which has to be inverted here is only a 3 × 3 matrix (or a 6 × 6 matrix in the case of a space frame) compared with the 6 × 6 (or 12 × 12) stiffness matrix appearing in (8.18). The final expressions for the member stress-resultants and the joint displacements are formally the same as (8.15) and (8.16).

As in the previous example, we may regard the release of joint *C* as an actual physical operation rather than a mathematical fiction. Thus we imagine that we start with a determinate structure *OABC*, which we load at *A* and *B*, leaving *C* entirely free. The effect of these loads is to produce a displacement at *C* of $\boldsymbol{u} = \mathbf{B}^t\mathbf{F_m B_0 p}$, where $\mathbf{B_0}$ is the relevant part of $\mathbf{H_+}^{-1}$ (dropping the last column since $\boldsymbol{q}$ is zero) and $\mathbf{B}^t$ is the relevant part of $(\mathbf{H_+}^{-1})^t$ (dropping the first two rows since only $\boldsymbol{u}$ is required).

We now consider the load $\boldsymbol{q}$ required to close the gap. This load is clearly one which, acting alone, would produce a displacement $-\boldsymbol{u}$ at *C*. If we think of *OABC* as a single member composed of three segments we may find its flexibility by using the analysis of Section 3.7. Since we have already assumed that the flexibility matrices of the segments *a*, *b* and *c* have been computed in the common coordinate system $Ox'y'$, the flexibility matrix of the composite member is $F_a + F_b + F_c$. Thus we have $e_{0c} = (F_a + F_b + F_c)r_{0c}$, and since $q = H_c'r_{0c}$, $e_{0c} = -H_c'^t u$. It follows that $u = -(H_c'^{-1})^t(F_a + F_b + F_c)H_c'^{-1}q$. In view of the form of $\mathbf{B}$ in (8.20) this is equivalent to $\boldsymbol{u} = -(\mathbf{B}^t\mathbf{F_m B})\boldsymbol{q}$, so that $\boldsymbol{q} = -(\mathbf{B}^t\mathbf{F_m B})^{-1}\boldsymbol{u}$. This is essentially the same result as (8.14).

Our decision to make the cut at C was of course quite arbitrary. We could equally well have chosen to make it at B, as shown in Fig. 8.5. With $\boldsymbol{q}$ defined as shown in the figure the augmented equilibrium equations are

$$\begin{bmatrix} p_A \\ p_B \\ q \end{bmatrix} = \begin{bmatrix} H_{2a}' & H_{1b}' & 0 \\ 0 & H_{2b}' & H_{1c}' \\ 0 & H_{2b}' & 0 \end{bmatrix} \begin{bmatrix} r_a \\ r_b \\ r_c \end{bmatrix}$$

Treating the frame as two separate tree structures we can write down the inverse equations as

$$\begin{bmatrix} r_a \\ r_b \\ r_c \end{bmatrix} = \left[\begin{array}{cc|c} (H_{2a}')^{-1} & 0 & (H_{2b}')^{-1} \\ 0 & 0 & (H_{2b}')^{-1} \\ 0 & (H_{1c}')^{-1} & -(H_{1c}')^{-1} \end{array}\right] \begin{bmatrix} p_A \\ p_B \\ q \end{bmatrix}$$

where, as before, the flexibility and H' matrices are referred to the common coordinate system $Ox'y'$. These equations may be written $\mathbf{r} = \bar{\mathbf{B}}_0\mathbf{p} + \bar{\mathbf{B}}\boldsymbol{q}$ and the analysis completed in the same way as before. Although the matrices $\bar{\mathbf{B}}_0$ and $\bar{\mathbf{B}}$ are different from $\mathbf{B}_0$ and $\mathbf{B}$, the computed values of $\mathbf{r}$ and $\mathbf{d}$ will be the same.

So far we have introduced all the releases at one point in the structure, allowing complete discontinuity at that point. This is the simplest way of making a rigid-jointed structure determinate, but it is by no means the only way. If our example is a plane frame we may achieve the same result by introducing three hinges at, say, A, B and C. The vector $\mathbf{q}$ now consists

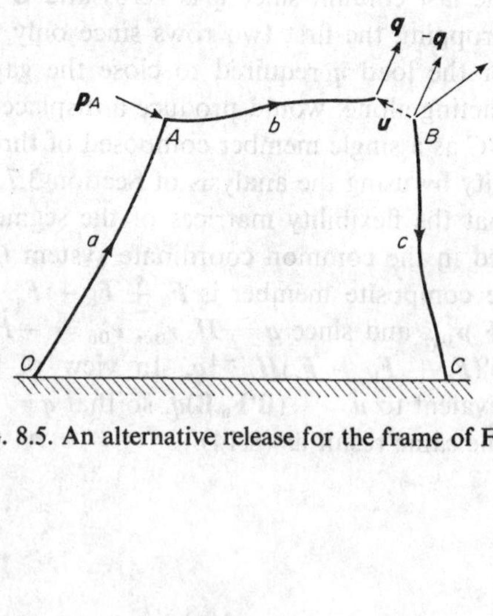

FIG. 8.5. An alternative release for the frame of Fig. 8.3.

of the moments at the three hinges, while the corresponding vector **u** consists of the three relative rotations.

The use of hinges rather than complete cuts in members has certain numerical advantages in the manual analysis of hyperstatic rigid-jointed frames. The matrix $\mathbf{B}^t\mathbf{F}_m\mathbf{B}$ is often better conditioned and the effect of inaccuracies in the calculated values of the release moments is less. These advantages, however, are obtained at the cost of additional complication in the inversion of the augmented equilibrium equations. The procedure for constructing $\mathbf{H}_+^{-1}$ for tree structures cannot be used, so that the inversion has to be done numerically, operating on individual scalar coefficients rather than on sub-matrices.

8.3. Applications to more complex skeletal structures

As we have seen, the compatibility method in its traditional form begins with the introduction of sufficient discontinuities (i.e. releases) into a structure to make it statically determinate. The number of releases required to make a structure determinate is known as the degree of indeterminacy of the structure. In most civil engineering structures this can be found by inspection, and for plane and space trusses with pin-joints there are well-known rules which give the degree of indeterminacy in terms of the number of members and the number of joints. For a general account of the problem of finding the degree of indeterminacy of a structure the reader is referred to Morice (1959) and Henderson and Bickley (1955).

It is relatively easy to calculate the degree of indeterminacy for frames in which all the joints are rigid. We merely introduce sufficient cuts in the members to convert the frame into a tree structure, which we know to be determinate. Each cut introduces three releases into a plane structure or six into a space structure, so that the degree of indeterminacy is simply three (or six) times the number of cuts. Alternatively we may argue that in the line diagram of a tree structure there are no closed paths or rings. A hyperstatic structure, on the other hand, always has a number of rings, each of which must be cut somewhere to make it into a tree structure. Hence the degree of indeterminacy is equal to three (or six) times the number of independent rings in the original structure. It may be noted here that the foundations must be included in the structure—if a rigid-jointed structure has two points of foundation attachment, then this constitutes a ring. The frame shown in Fig. 8.6a, for example, has four independent rings, and has therefore 12 or 24 degrees of indeterminacy, according as to whether it is treated as a plane or a space frame. It may be converted into a tree structure by the introduction of four cuts, whose positions may (with some restrictions†) be chosen arbitrarily. Two possible systems of releases are shown in Figs. 8.6b and 8.6c.

A more difficult type of structure to deal with is one with some joints rigid and some joints pinned. One way of tackling such a structure is to temporarily make all the pin-joints rigid, and find the degree of indeterminacy of the resulting structure by counting the number of independent rings in the manner described above. If we subtract from this number the number of additional constraints introduced by making the pin-joints rigid we obtain the degree of indeterminacy of the original structure.

While it is relatively easy to determine the number of releases required to make a given structure determinate, it is difficult to give rigid rules for choosing the positions of the releases. The characteristics of a good set of releases are:

(a) The numerical work should be as simple as possible. In other words it should be easy to find the matrices $\mathbf{B}_0$ and $\mathbf{B}$, and these matrices should if possible contain a large number of zeros.

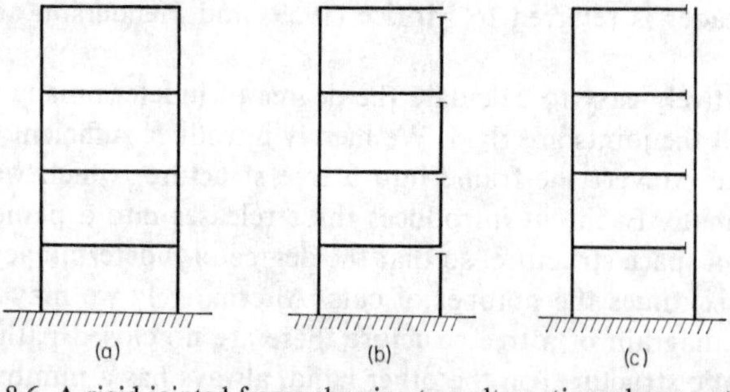

FIG. 8.6. A rigid-jointed frame showing two alternative release systems.

† The restriction on the positions of the cuts is that they must break all the rings—otherwise they will reduce the structure to a combination of a hyperstatic structure and one or more sections with rigid-body degrees of freedom.

(b) The matrix $\mathbf{B}^t\mathbf{F}_m\mathbf{B}$ should be well conditioned, in order to avoid trouble during the inversion process.

(c) The system of internal forces and moments induced by the release loads $\mathbf{q}$ (the system $\mathbf{Bq}$ in (8.24) below) should be small compared with the forces and moments induced in the released structure by the applied loads (the system $\mathbf{B}_0\mathbf{p}$). This reduces the effects of inaccuracies in the calculation of $\mathbf{q}$.

Having decided on the number and positions of the releases the analysis is very similar to that presented in Section 8.2. We add the self-equilibrating load-pairs $\pm q_1$, $\pm q_2$, . . ., associated with the releases to the applied load vector, at the same time adding appropriate rows to the matrix $\mathbf{H}$. Thus we obtain an equation which we write

$$\begin{bmatrix}\mathbf{p}\\ \mathbf{q}\end{bmatrix} = \mathbf{H}_+\mathbf{r} \tag{8.23}$$

Inverting $\mathbf{H}_+$ we obtain the expression

$$\mathbf{r} = \mathbf{H}_+^{-1}\begin{bmatrix}\mathbf{p}\\ \mathbf{q}\end{bmatrix}$$

which we write as

$$\mathbf{r} = \mathbf{B}_0\mathbf{p} + \mathbf{Bq}. \tag{8.24}$$

Alternatively we may construct the matrices $\mathbf{B}_0$ and $\mathbf{B}$ directly, using the rules given in Section 6.4. The matrices $\mathbf{B}_0$ and $\mathbf{B}$ have as many rows as $\mathbf{r}$ has components. $\mathbf{B}_0$ has a column for each component of the external load vector $\mathbf{p}$, and $\mathbf{B}$ has a column for each component of $\mathbf{q}$. The product $\mathbf{B}_0\mathbf{p}$ gives the internal load system produced in the structure when all the release load-pairs are zero, while the product $\mathbf{Bq}$ represents the self-equilibrating internal load system corresponding to the release load-pairs alone.

Corresponding to (8.24) we have the relationship between displacements and deformations, which we write as

$$\begin{bmatrix}\mathbf{d}\\ \mathbf{u}\end{bmatrix} = \begin{bmatrix}\mathbf{B}_0{}^t\\ \mathbf{B}^t\end{bmatrix}\mathbf{e} \tag{8.25}$$

or as $\mathbf{d} = \mathbf{B}_0{}^t\mathbf{e}$, $\mathbf{u} = \mathbf{B}^t\mathbf{e}$. The displacement vector $\mathbf{d}$ represents the displacements of the points of application of the known external loads, while $\mathbf{u}$ represents the *relative* displacements at the releases.

We now impose the condition that these relative displacements must be zero. This gives

$$\mathbf{u} = \mathbf{B}^t\mathbf{e} = \mathbf{B}^t\mathbf{F}_m\mathbf{r} = \mathbf{B}^t\mathbf{F}_m(\mathbf{B}_0\mathbf{p} + \mathbf{B}\mathbf{q}) = \mathbf{0}. \tag{8.26}$$

Thus we obtain the value of $\mathbf{q}$ as

$$\mathbf{q} = -(\mathbf{B}^t\mathbf{F}_m\mathbf{B})^{-1}\mathbf{B}\mathbf{F}_m\mathbf{B}_0\mathbf{p}. \tag{8.27}$$

Substituting this expression for $\mathbf{q}$ in (8.24) we find the true internal loads $\mathbf{r}$, and the displacements follow from the equation $\mathbf{d} = \mathbf{B}_0{}^t\mathbf{e} = \mathbf{B}_0{}^t\mathbf{F}_m\mathbf{r}$.

An interesting feature of the compatibility method is the way in which the matrix $\mathbf{B}^t\mathbf{F}_m\mathbf{B}$, which is the flexibility matrix of the structure as measured at the releases, resembles the matrix $\mathbf{H}\mathbf{K}_m\mathbf{H}^t$, which is the stiffness matrix of the structure as measured at the joints. Consider, for example, the release system shown in Fig. 8.6b. Here we have four independent rings, with release displacements u_1, u_2, u_3, u_4 and self-equilibrating release load-pairs $\pm q_1$, $\pm q_2$, $\pm q_3$, $\pm q_4$ as shown in Fig. 8.7.

The u's are expressed in terms of the q's by the linear equation $\mathbf{u} = \mathbf{B}^t\mathbf{F}_m\mathbf{B}\mathbf{q}$, which in this case may be written as

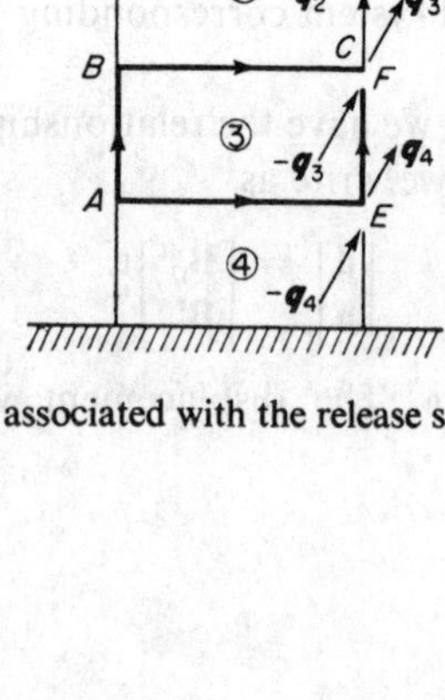

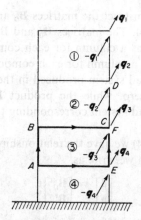

FIG. 8.7. The load-pairs associated with the release system shown in Fig. 8.6b.

$$\begin{bmatrix} u_1 \\ u_2 \\ u_3 \\ u_4 \end{bmatrix} = \begin{bmatrix} F_{11} & F_{12} & F_{13} & F_{14} \\ F_{21} & F_{22} & F_{23} & F_{24} \\ F_{31} & F_{32} & F_{33} & F_{34} \\ F_{41} & F_{42} & F_{43} & F_{44} \end{bmatrix} \begin{bmatrix} q_1 \\ q_2 \\ q_3 \\ q_4 \end{bmatrix} \tag{8.28}$$

We now consider the detailed form of the matrix appearing in (8.28). In particular we direct our attention to the third row, whose coefficients give us the relative displacement u_3 of the points C and F due to the various q's. The leading diagonal sub-matrix F_{33} gives us the displacement due to the load-pair q_3 applied at C and F. The only members affected by this load-pair are the members comprising the ring $FEABC$, and the flexibility matrix of this ring may easily be found from the analysis in Section 3.7 as

$$F_{33} = F_{AB}' + F_{BC}' + F_{AE}' + F_{EF}'$$

where the flexibility matrices F_{AB}', etc., are computed with C (or F) as origin. Applying the same process to the other rings we see that in general each leading diagonal element of the flexibility matrix $\mathbf{B}^t\mathbf{F}_m\mathbf{B}$ is equal to the sum of the *flexibilities* of the members making up the corresponding ring. This may be compared with the fact that in the stiffness matrix $\mathbf{HK}_m\mathbf{H}^t$ of a skeletal structure each leading diagonal element is the sum of the *stiffnesses* of the members meeting at the corresponding joint.

If we now look at the off-diagonal elements of the third row, we see that the relative displacement of the points C and F due to the load-pair q_2 is caused entirely by the strain induced in the member BC, this member being the only one common to rings 2 and 3. Hence F_{32} is a function of F_{BC}' only. Similarly the contribution to u_3 from the load-pair q_4 is caused entirely by the strain in member AE, this being the only member common to rings 3 and 4, so that F_{34} is a function of F_{AE}' only. Finally we see that F_{31} must be zero. For rings 1 and 3 have no common member, and therefore no strains are caused in the members of ring 3 by any system of loads in ring 1. Applying the same analysis to the other rings we see that in general an off-diagonal element in the flexibility matrix depends on the flexibility of the member (or members) common to the two corresponding rings. If the rings have no members in common then the element is zero. This may be compared with the fact that in the stiffness matrix of a structure an off-diagonal element is a function of the stiffness of the member connecting the two corresponding joints, and is zero if the joints are not directly connected by a member. It follows that for the release system shown in Fig. 8.6b equation (8.28) is actually of the form

$$\begin{bmatrix} \boldsymbol{u}_1 \\ \boldsymbol{u}_2 \\ \boldsymbol{u}_3 \\ \boldsymbol{u}_4 \end{bmatrix} = \begin{bmatrix} \boldsymbol{F}_{11} & \boldsymbol{F}_{12} & \boldsymbol{0} & \boldsymbol{0} \\ \boldsymbol{F}_{21} & \boldsymbol{F}_{22} & \boldsymbol{F}_{23} & \boldsymbol{0} \\ \boldsymbol{0} & \boldsymbol{F}_{32} & \boldsymbol{F}_{33} & \boldsymbol{F}_{34} \\ \boldsymbol{0} & \boldsymbol{0} & \boldsymbol{F}_{43} & \boldsymbol{F}_{44} \end{bmatrix} \begin{bmatrix} \boldsymbol{q}_1 \\ \boldsymbol{q}_2 \\ \boldsymbol{q}_3 \\ \boldsymbol{q}_4 \end{bmatrix}$$

Like the stiffness matrix of a structure, the flexibility matrix tends to be "banded" in form, having its elements grouped round the leading diagonal. As we shall see in Section 11.1, this feature has considerable computational advantages. However, the banded nature of the flexibility matrix depends on the arrangement of the rings, and this in turn depends on the positions chosen for the releases. For example, if we had chosen the system of releases shown in Fig. 8.6c the four rings would be as shown in Fig.8.8. A load applied at one release point would produce displacements at all the others, so that there would be no zero sub-matrices in the flexibility matrix. It follows that it is desirable to choose release systems in such a way that the rings are as independent of each other as possible.

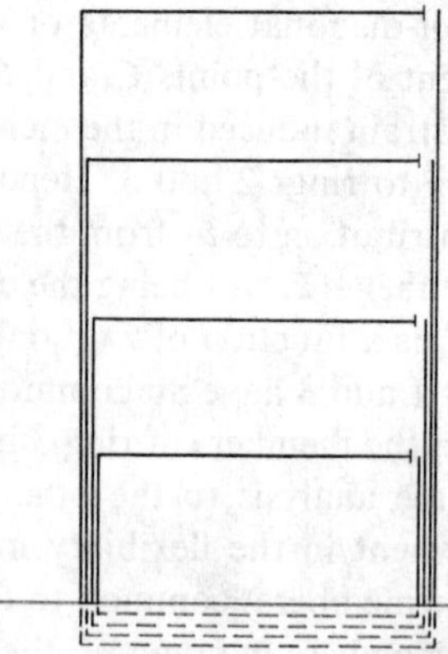

FIG. 8.8. The system of rings associated with the releases shown in Fig. 8.6c.

8.4. Axial forces in rigid-jointed frames

Because of its systematic nature the equilibrium method described in Chapter 5 is easily programmed for a computer. The systematic nature of the analysis is due largely to the fact that all the joints in a given structure are treated as having the same number of degrees of freedom—two per joint in a pin-jointed plane truss, three per joint in a rigid-jointed plane frame, etc. This allows the same method of analysis (and, by implication, the same computer program) to be applied to many different types of structure. One result of this approach is that the equilibrium method always includes the effects of axial strains in members, making no distinction between frames and trusses.

While it is useful to have a general method of analysis which is "exact" in the sense of including the axial strain effects, it is also desirable to have an equally systematic technique which neglects these effects, and therefore involves less computation. Unfortunately it is difficult to do this in the matrix equilibrium method without destroying the systematic nature of the method. It is computationally impossible to make the axial stiffness of a member infinite, and merely making it large achieves nothing in the way of simplification. In fact, as we shall see in Section 11.4, this may lead to ill-conditioning of the load/displacement equations.

In the compatibility method, however, some simplification is possible. We cannot neglect axial forces entirely, since an axial force in one member may cause a shear force to appear in another. Thus in writing the equations of joint equilibrium $\mathbf{p} = \mathbf{Hr}$, we must include axial forces in the vector of member stress-resultants $\mathbf{r}$. However, when the matrix $\mathbf{H}$ has been augmented and inverted we may drop the axial force components from $\mathbf{r}$, keeping the values of the other components unchanged. Hence we may write

$$\mathbf{r} = \mathbf{B}_0\mathbf{p} + \mathbf{Bq} \tag{8.29}$$

where $\mathbf{r}$ has no axial force components, and the corresponding rows have been omitted from the two matrices $\mathbf{B}_0$ and $\mathbf{B}$.

While an infinite stiffness is a computational embarrassment, a zero flexibility is easily handled. Having dropped the axial force component from each member stress-resultant appearing in $\mathbf{r}$ it is natural to drop the axial strain component from each member deformation appearing in $\mathbf{e}$.

Thus each member flexibility matrix becomes a 2×2 matrix in the case of a plane frame, or a 5×5 matrix in the case of a space frame. It is clear that this procedure does not affect the expressions for the other components of $\mathbf{e}$—it merely eliminates the calculations associated with the axial strain components, which we are assuming to be zero.

The expressions for the displacements at the joints and the releases are still $\mathbf{d} = \mathbf{B}_0{}^t\mathbf{e}$, $\mathbf{u} = \mathbf{B}^t\mathbf{e}$, where $\mathbf{B}_0$ and $\mathbf{B}$ are the reduced matrices appearing in equation (8.29). Dropping the rows from $\mathbf{B}_0$ and $\mathbf{B}$ which are associated with axial forces eliminates the corresponding columns from $\mathbf{B}_0{}^t$ and $\mathbf{B}^t$, and it is precisely these columns which correspond to the zero axial strains which we have omitted from $\mathbf{e}$. The remainder of the analysis is formally the same as that given in the previous section. In the case of a plane frame the overall result is a reduction of a third in the number of rows in the matrices $\mathbf{B}_0$ and $\mathbf{B}$, and a reduction of over 50% in the number of coefficients in the member flexibility matrices. A similar technique may be employed in cases where a member can be regarded as infinitely rigid in bending.

8.5. An algebraic procedure for transforming the equilibrium equations

We have seen that the first part of the compatibility method is concerned with the transformation of the nodal equilibrium equations $\mathbf{p} = \mathbf{Hr}$ into the form $\mathbf{r} = \mathbf{B}_0\mathbf{p} + \mathbf{Bq}$. So far we have selected the vector of "redundants" $\mathbf{q}$ by reasoning based on the topology of the structure being analysed. We now consider the transformation from a purely algebraic point of view. The procedure we shall describe is an extension of the Gauss–Jordan procedure for matrix inversion described in Section 6 of the Appendix. It has been termed the "rank technique" by Robinson (1966). The first part is almost identical to the first part of the procedure for collapse load-factor analysis described in Section 7.7.

We begin by writing the equilibrium equations $\mathbf{p} = \mathbf{Hr}$ in the form $\mathbf{Ip} = \mathbf{Hr}$, where $\mathbf{I}$ is a unit matrix (see Fig. 8.9). We now commence the Gauss–Jordan procedure described in Section 6 of the Appendix just as though the matrix $\mathbf{H}$ were square. The sequence is:

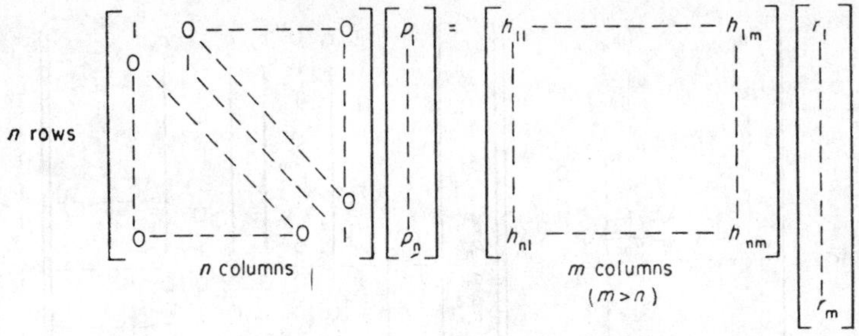

FIG. 8.9.

"For each equation $i = 1, \ldots, n$

(a) Find the coefficient in row i of **H** which has largest modulus. Let this be in column j.

(b) Carry out a Gauss–Jordan step using h_{ij} as pivot."

Since there are only n equations this process only reduces n of the m columns of **H**, leaving the equations in a form such as

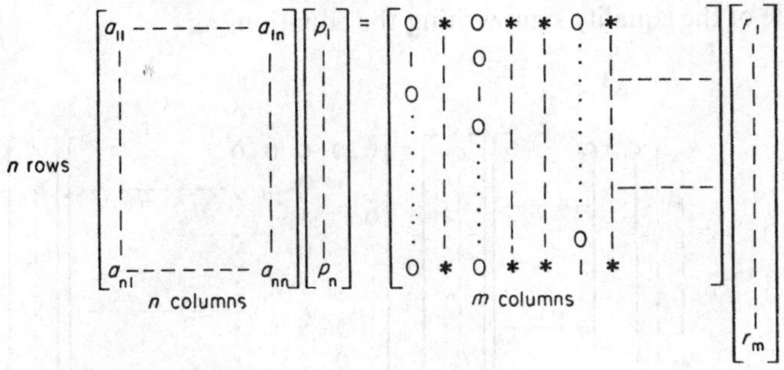

The *'s indicate the m-n columns from which pivots have *not* been selected during the reduction process. Let these columns be denoted by $\mathbf{c}_1, \mathbf{c}_2, \ldots, \mathbf{c}_{m-n}$. The components of **r** associated with these columns are chosen to form the vector **q**. We therefore augment the existing equations with equations identifying these components.

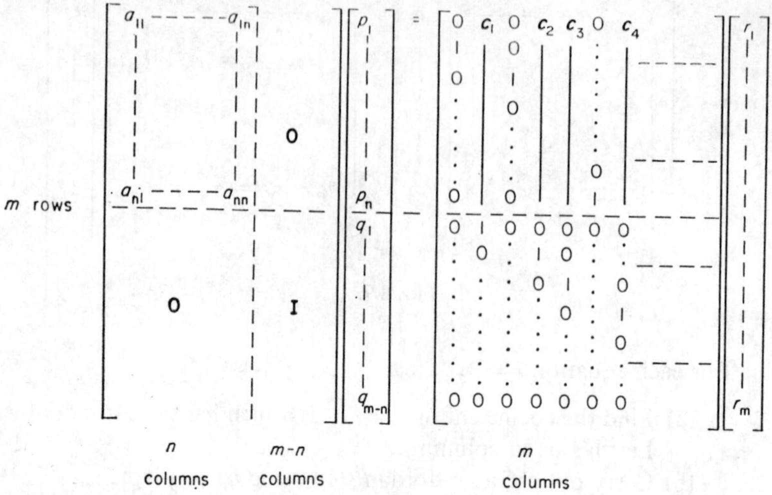

Since the columns $c_1, c_2, \ldots, c_{m-n}$ multiply components of $\mathbf{r}$ which are identical to $q_1, q_2, \ldots, q_{m-n}$ we may transfer these columns to the left-hand side of the equality sign, writing the equations as

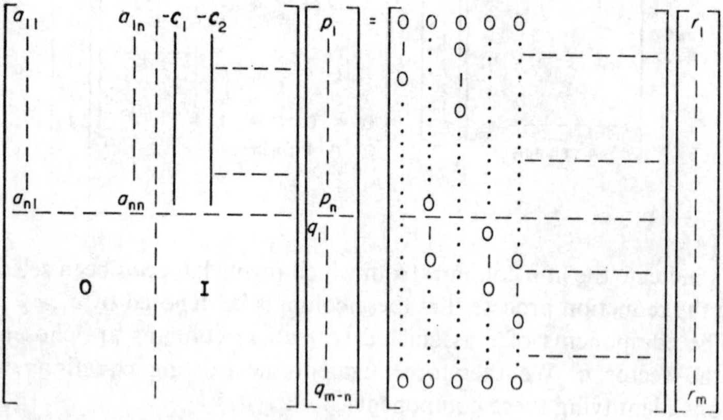

The right-hand matrix is now a matrix with a single 1 in each row and column. As in the normal inversion procedure, we may convert it to a unit matrix by appropriate rearrangement of the equations, giving

$$m \text{ rows} \left[\begin{array}{c|c} \mathbf{B}_0 & \mathbf{B} \\ \underset{\text{columns}}{n} & \underset{\text{columns}}{m-n} \end{array} \right] \begin{bmatrix} \mathbf{p} \\ \mathbf{q} \end{bmatrix} = \mathbf{r}$$

which is the required equation

$$\mathbf{r} = \mathbf{B}_0\mathbf{p} + \mathbf{B}\mathbf{q}.$$

There are a number of variations on this numerical procedure. For further details and examples of the method the reader is referred to the text by Robinson mentioned earlier.

Having reached this point in the analysis without introducing "releases" it is natural to see whether we can derive the remaining equations of the compatibility method without having to consider the closing of fictitious "gaps". We can do this by a simple argument based on virtual work.

Let $\mathbf{p}^*$, $\mathbf{q}^*$ and $\mathbf{r}^*$ be any set of *virtual* loads and stress-resultants satisfying the equilibrium equations $\mathbf{r}^* = \mathbf{B}_0\mathbf{p}^* + \mathbf{B}\mathbf{q}^*$, and let $\mathbf{d}$ and $\mathbf{e}$ be the nodal displacement and member deformation vectors produced in the structure by the *actual* applied loads. The virtual work equation linking these two sets of vectors is

$$\mathbf{p}^{*t}\mathbf{d} = \mathbf{r}^{*t}\mathbf{e} = (\mathbf{p}^{*t}\mathbf{B}_0{}^t + \mathbf{q}^{*t}\mathbf{B}^t)\mathbf{e}.$$

In this equation $\mathbf{p}^*$ and $\mathbf{q}^*$ may be chosen arbitrarily and independently. It follows that $\mathbf{d} = \mathbf{B}_0{}^t\mathbf{e}$ and $\mathbf{B}^t\mathbf{e} = \mathbf{0}$. The first of these equations agrees with (8.25), while the second is identical with the condition (8.26), which was derived by making the release discontinuities $\mathbf{u}$ equal to zero. The rest of the analysis now follows on as before.

The analysis in this section shows that it is possible to view the compatibility method as an algebraic process involving only the matrices $\boldsymbol{H}'$

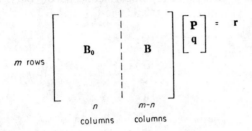

and $\boldsymbol{F}$. This approach makes it possible to compare it with the equilibrium method, which involves only the matrices $\boldsymbol{H}'$ and $\boldsymbol{K}$. The formal approach is particularly useful in applications of the compatibility method to finite element problems, where the topological concepts of "cuts" and "releases" are difficult to visualize.

CHAPTER 9

Transfer Matrices

So far in this book we have been concerned with methods of analysis which can be applied to structures of arbitrary geometrical form. As we have seen, these methods involve the manipulation of matrices whose sizes depend on the number of members or the number of joints in a structure. We now develop a technique for analysing a particular class of structure which does not require the construction and manipulation of large matrices.

The technique can be applied to any structure in which the joints are connected by a single chain of members. Examples of such structures are shown in Figs. 9.3 and 9.4—their essential characteristic is that if the joints are numbered sequentially then joint J is connected to joints $J+1$ and $J-1$, and possibly to a rigid foundation, but to no other joints. The idea may be extended to cover structures, such as lattice towers, which can be treated as a number of parallel chains. Further information about the method may be found in the text by Pestel and Leckie (1963).

9.1. Transfer matrices for single structural elements

Figure 9.1 shows part of a chain of structural elements in which the joints are numbered sequentially from left to right. Let $\boldsymbol{d}_X$ be the displacement of a point X and let $\boldsymbol{p}_X$ be the load-vector acting on the *left-hand side* of a section through the chain at X. (This convention for $\boldsymbol{p}_X$, while arbitrary, agrees with that used in Fig. 3.1b.) We call the vector

$$\begin{bmatrix} \boldsymbol{d}_X \\ \boldsymbol{p}_X \end{bmatrix}$$

the *state vector* at the point X and denote it by $\boldsymbol{s}_X$. It follows from this

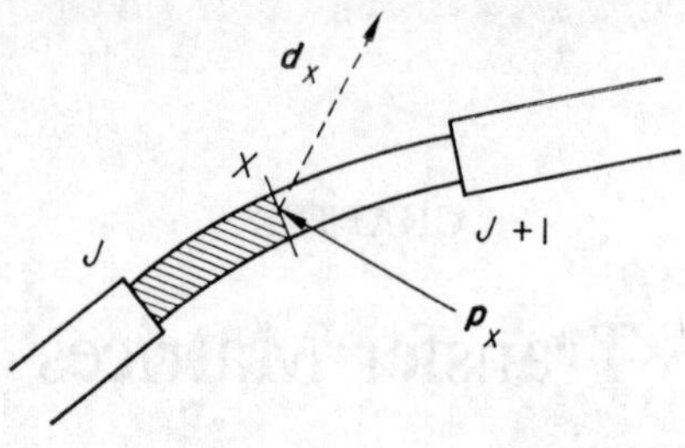

FIG. 9.1. A typical element in a chain of members.

definition that s has 6 components in a plane structure and 12 in a three-dimensional one. State vectors may be expressed in either a local or a global coordinate system—transformation from one system to the other is simply a matter of operating on the individual p and d components in the usual way.

The method of transfer matrices involves the calculation of s at successive points of the chain, starting (conventionally) at the left-hand end. This requires, for each member and joint, an expression which gives the state vector at the right-hand end in terms of the state vector at the left-hand end.

9.1.1. *The equations for a line element*

It is easy to see from Fig. 9.1 that the state vector at end 2 of an element is

$$s_2 = \begin{bmatrix} d_2 \\ p_2 \end{bmatrix},$$

while the state vector at end 1 is

$$s_1 = \begin{bmatrix} d_1 \\ -p_1 \end{bmatrix},$$

where p_1 and p_2 are the loads acting on the ends of the element. The minus sign appears in the expression for s_1 because the load component of s always follows a "right acting on left" convention, while the definition of p_1 in Chapter 3 implies a "left acting on right" convention.

To obtain s_2 in terms of s_1 we start from the load/displacement equations of a line element, as given in (3.10a)

$$\left.\begin{aligned} p_1 &= H_1 K H_1{}^t d_1 + H_1 K H_2{}^t d_2, \\ p_2 &= H_2 K H_1{}^t d_1 + H_2 K H_2{}^t d_2. \end{aligned}\right\} \tag{9.1}$$

The first equation of (9.1) may be written in the form

$$d_2 = -(H_1 K H_2{}^t)^{-1} H_1 K H_1{}^t d_1 + (H_1 K H_2{}^t)^{-1} p_1$$

and since $(H_1 K H_2{}^t)^{-1} = (H_2{}^t)^{-1} K^{-1} H_1{}^{-1}$ this becomes

$$d_2 = -(H_2{}^t)^{-1} H_1{}^t d_1 + K_{12}{}^{-1} p_1. \tag{9.2}$$

It may also be deduced from (9.1) that

$$p_2 = H_2 H_1{}^{-1} p_1. \tag{9.3}$$

Evaluating the product $H_2 H_1{}^{-1}$, or directly from static equilibrium considerations, we obtain

$$-H_2 H_1{}^{-1} = \begin{bmatrix} 1 & 0 & 0 \\ 0 & 1 & 0 \\ (y_2 - y_1) & -(x_2 - x_1) & 1 \end{bmatrix}$$

Following the notation introduced in equation (3.32), we write this matrix as H_{21}. Evaluating the product $-(H_2{}^t)^{-1} H_1{}^t$ appearing in (9.2) we obtain

$$-(H_2{}^t)^{-1} H_1{}^t = -(H_1 H_2{}^{-1})^t = \begin{bmatrix} 1 & 0 & -(y_2 - y_1) \\ 0 & 1 & (x_2 - x_1) \\ 0 & 0 & 1 \end{bmatrix}$$

which we write as $H_{12}{}^t$. Using this notation equations (9.2) and (9.3) may be combined and written in the form

$$\begin{bmatrix} d_2 \\ p_2 \end{bmatrix} = \begin{bmatrix} H_{12}{}^t & -K_{12}{}^{-1} \\ 0 & H_{21} \end{bmatrix} \begin{bmatrix} d_1 \\ -p_1 \end{bmatrix} \tag{9.4}$$

This matrix equation is completely equivalent to (9.1). The vectors appearing in (9.4) are the state vectors s_1, s_2 and the matrix which connects them is defined as the *transfer matrix* of the member. We write (9.4) as

$$\boldsymbol{s}_2 = \boldsymbol{G}\boldsymbol{s}_1. \tag{9.5}$$

Interchanging suffixes in (9.4) and making suitable changes of sign we obtain

$$\begin{bmatrix} \boldsymbol{d}_1 \\ -\boldsymbol{p}_1 \end{bmatrix} = \begin{bmatrix} \boldsymbol{H}_{21}{}^t & \boldsymbol{K}_{21}{}^{-1} \\ \boldsymbol{0} & \boldsymbol{H}_{12} \end{bmatrix} \begin{bmatrix} \boldsymbol{d}_2 \\ \boldsymbol{p}_2 \end{bmatrix}$$

which we write as

$$\boldsymbol{s}_1 = \boldsymbol{G}^{-1}\boldsymbol{s}_2. \tag{9.6}$$

For a uniform straight member the value of $\boldsymbol{K}_{12}{}^{-1}$ may be obtained from (3.11) as

$$\boldsymbol{K}_{12}{}^{-1} = \begin{bmatrix} -L/EA & 0 & 0 \\ 0 & L^3/6EI & -L^2/2EI \\ 0 & L^2/2EI & -L/EI \end{bmatrix}$$

It follows that for such a member

$$\boldsymbol{G} = \begin{bmatrix} 1 & 0 & 0 & L/EA & 0 & 0 \\ 0 & 1 & L & 0 & -L^3/6EI & L^2/2EI \\ 0 & 0 & 1 & 0 & -L^2/2EI & L/EI \\ 0 & 0 & 0 & 1 & 0 & 0 \\ 0 & 0 & 0 & 0 & 1 & 0 \\ 0 & 0 & 0 & 0 & -L & 1 \end{bmatrix}$$

In this analysis we have assumed that any loads acting at internal points of the element have already been replaced by equivalent loads acting on the joints at its ends. As in Section 5.3, it is possible to incorporate these loads in (9.4) by starting from the load/displacement equations

$$\boldsymbol{p}_1 = \boldsymbol{H}_1\boldsymbol{K}\boldsymbol{H}_1{}^t\boldsymbol{d}_1 + \boldsymbol{H}_1\boldsymbol{K}\boldsymbol{H}_2{}^t\boldsymbol{d}_2 - (\boldsymbol{p}_{\text{equiv}})_1,$$

$$\boldsymbol{p}_2 = \boldsymbol{H}_2\boldsymbol{K}\boldsymbol{H}_1{}^t\boldsymbol{d}_1 + \boldsymbol{H}_2\boldsymbol{K}\boldsymbol{H}_2{}^t\boldsymbol{d}_2 - (\boldsymbol{p}_{\text{equiv}})_2$$

rather than from equations (9.1).

Transformation of (9.4) into global coordinates follows the normal rules. The matrix $\boldsymbol{K}_{12}{}^{-1}$ is replaced by $(\boldsymbol{K}_{12}')^{-1}$ and the coordinates x_1, y_1, x_2, y_2 in the products $\boldsymbol{H}_{12}{}^t$ and $\boldsymbol{H}_{21}$ are replaced by their corresponding global values. The resulting equations are written as

$$s_2' = G's_1'. \tag{9.7}$$

The equations for a general three-dimensional element are formally the same as those given above, provided that the H and K matrices are defined appropriately and the necessary components are included in s_1 and s_2.

9.1.2. *The equations for a joint*

A joint is simply a point where two members meet. We denote the two sides of the joint by the subscripts $X-$ (left-hand side) and $X+$ (right-hand side). We consider the four basic cases shown in Fig. 9.2.

1. The joint is rigid and there is no foundation connection or applied load. The compatibility condition at the joint is $d_{X+} = d_X = d_{X-}$ and the equilibrium condition is $p_{X+} = p_{X-}$. (Note that our convention for p implies that p_{X+} is the load exerted on the joint *by* the member to its right, while p_{X-} is the load which the joint exerts *on* the member to its left.) Thus $s_{X+} = s_{X-}$ and the transfer matrix associated with the joint is simply a unit matrix.
2. The joint is rigid, there is no foundation connection, but there is an applied load p_X. The equilibrium equation is now $p_{X+} + p_X = p_{X-}$

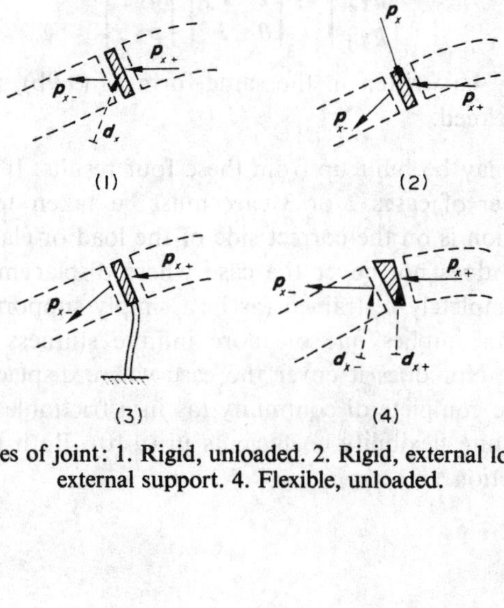

FIG. 9.2. Types of joint: 1. Rigid, unloaded. 2. Rigid, external load. 3. Rigid, external support. 4. Flexible, unloaded.

and the relation between the state vectors is

$$\begin{bmatrix} \boldsymbol{d}_{X+} \\ \boldsymbol{p}_{X+} \end{bmatrix} = \begin{bmatrix} \boldsymbol{d}_{X-} \\ \boldsymbol{p}_{X-} \end{bmatrix} - \begin{bmatrix} \boldsymbol{0} \\ \boldsymbol{p}_{X} \end{bmatrix} \tag{9.8a}$$

This may be written as

$$\boldsymbol{s}_{X+} = \boldsymbol{s}_{X-} - \boldsymbol{s}_{X}. \tag{9.8b}$$

3. The joint is rigid, unloaded and is connected elastically to a foundation. The equilibrium equation is now $\boldsymbol{p}_{X+} = \boldsymbol{p}_{X-} + \boldsymbol{K}_X\boldsymbol{d}_X$, where $\boldsymbol{K}_X$ is the stiffness matrix of the support as measured at the joint. Since $\boldsymbol{d}_{X+} = \boldsymbol{d}_X = \boldsymbol{d}_{X-}$ this may be written as

$$\begin{bmatrix} \boldsymbol{d}_{X+} \\ \boldsymbol{p}_{X+} \end{bmatrix} = \begin{bmatrix} \boldsymbol{I} & \boldsymbol{0} \\ \boldsymbol{K}_X & \boldsymbol{I} \end{bmatrix} \begin{bmatrix} \boldsymbol{d}_{X-} \\ \boldsymbol{p}_{X-} \end{bmatrix} \tag{9.9a}$$

or as

$$\boldsymbol{s}_{X+} = \boldsymbol{G}_X\boldsymbol{s}_{X-}. \tag{9.9b}$$

4. The joint is flexible and there is no applied load or foundation support. The equilibrium condition is $\boldsymbol{p}_{X+} = \boldsymbol{p}_{X-}$, as in case 1, but the compatibility condition is now $\boldsymbol{d}_{X+} = \boldsymbol{d}_{X-} + \boldsymbol{F}_X\boldsymbol{p}_{X-}$, where $\boldsymbol{F}_X$ is the flexibility matrix of the connection. Thus we have

$$\begin{bmatrix} \boldsymbol{d}_{X+} \\ \boldsymbol{p}_{X+} \end{bmatrix} = \begin{bmatrix} \boldsymbol{I} & \boldsymbol{F}_X \\ \boldsymbol{0} & \boldsymbol{I} \end{bmatrix} \begin{bmatrix} \boldsymbol{d}_{X-} \\ \boldsymbol{p}_{X-} \end{bmatrix} \tag{9.10}$$

which may be written in the same form as (9.9b), provided $\boldsymbol{G}_X$ is suitably defined.

Other cases may be built up from these four results. If case 4 is combined with either of cases 2 or 3 care must be taken to see that the flexible connection is on the correct side of the load or elastic support.

This analysis does not cover the case where displacement degrees of freedom are completely restrained (as in a simply supported continuous beam), since this implies one or more infinite stiffness coefficients in equation (9.9a). Nor does it cover the case where displacement degrees of freedom have complete discontinuity (as in a frictionless hinge), since this implies infinite flexibility coefficients in (9.10). Both these cases are discussed in Section 9.5.

9.2. An example of the use of transfer matrices

For our first example of the use of transfer matrices we consider the built-in arch shown in Fig. 9.3. We imagine that the transfer matrix for

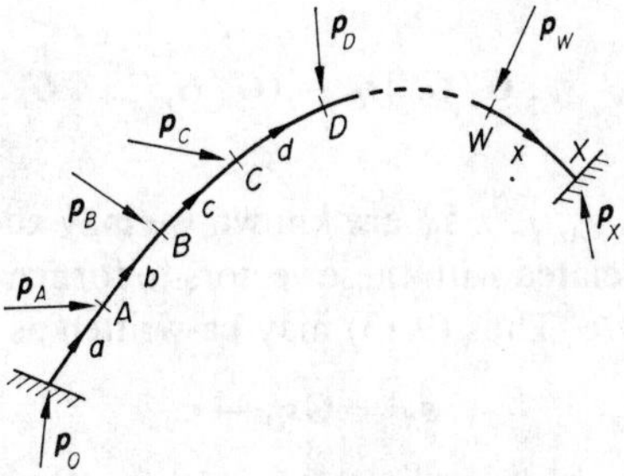

FIG. 9.3. An example of a built-in arch.

each section of the arch is known in global coordinates—as we have seen, this merely implies that we know the $\boldsymbol{K}$, $\boldsymbol{H}$ and $\boldsymbol{T}$ matrices associated with the sections. The loads $\boldsymbol{p}_A, \boldsymbol{p}_B, \ldots$ are assumed to be known and to include any equivalent joint loads due to distributed forces acting on the members.

At the left-hand end of section a we have the state vector

$$\boldsymbol{s}_{1a}' = \begin{bmatrix} \boldsymbol{d}_{1a}' \\ -\boldsymbol{p}_{1a} \end{bmatrix} = \begin{bmatrix} \boldsymbol{0} \\ -\boldsymbol{p}_O \end{bmatrix} = \boldsymbol{s}_O$$

where $\boldsymbol{p}_O$ is the unknown reaction at the support O. Multiplying $\boldsymbol{s}_{1a}'$ by the transfer matrix for member a we obtain the state vector just to the left of joint A as

$$\boldsymbol{s}_{2a}' = \boldsymbol{s}_{A-} = \boldsymbol{G}_a'\boldsymbol{s}_{1a}' = \boldsymbol{G}_a'\boldsymbol{s}_O. \tag{9.11}$$

The state vector to the right of joint A is now given by (9.8b) as

$$\boldsymbol{s}_{A+} = \boldsymbol{s}_{1b}' = \boldsymbol{s}_{A-} - \boldsymbol{s}_A \tag{9.12}$$

where $\boldsymbol{s}_A = \begin{bmatrix} \boldsymbol{0} \\ \boldsymbol{p}_A \end{bmatrix}$ specifies the external load at A.

Combining (9.11) and (9.12) gives

$$\boldsymbol{s}_{1b}' = \boldsymbol{G}_a'\boldsymbol{s}_O - \boldsymbol{s}_A.$$

and repeating the process for member *b* and joint *B* we obtain

$$s_{1c}' = G_b'G_a's_O - G_b's_A - s_B.$$

Progressing in this manner along the whole length of the arch we finally obtain

$$s_{2x}' = s_X = (G_x'G_w' \ldots G_b'G_a')s_O - (G_x'G_w' \ldots G_b')s_A - \ldots - G_x's_W. \tag{9.13}$$

Since all the vectors $s_A, \ldots, s_W$ are known we may add up all the terms in equation (9.13) associated with these vectors to form a single known vector which we denote by **c**. Thus (9.13) may be written as

$$s_X = \mathbf{G}s_O - \mathbf{c}. \tag{9.14}$$

It is clear that although the matrix **G** and the vector **c** relate to the whole arch they are the same size as the individual transfer matrices and state vectors associated with the separate sections.

We have already introduced the condition of zero displacement at end *O* by writing

$$s_O = \begin{bmatrix} 0 \\ -p_O \end{bmatrix}$$

We now introduce the corresponding condition at end *X* by writing

$$s_X = \begin{bmatrix} 0 \\ p_X \end{bmatrix}$$

If we split the matrix **G** into sub-matrices $\mathbf{G}_{11}$, $\mathbf{G}_{12}$, etc., we may write equation (9.14) in the form

$$\begin{bmatrix} 0 \\ p_X \end{bmatrix} = \begin{bmatrix} \mathbf{G}_{11} & \mathbf{G}_{12} \\ \mathbf{0} & \mathbf{G}_{22} \end{bmatrix} \begin{bmatrix} 0 \\ -p_O \end{bmatrix} - \begin{bmatrix} \mathbf{c}_1 \\ \mathbf{c}_2 \end{bmatrix}$$

or as

$$0 = -\mathbf{G}_{12}p_O - \mathbf{c}_1 \tag{9.15}$$

$$p_X = -\mathbf{G}_{22}p_O - \mathbf{c}_2 \tag{9.16}$$

Equation (9.15) may now be solved for the unknown support reaction p_O, after which equation (9.16) (which is merely the overall equilibrium equation for the arch) gives the other reaction $\boldsymbol{p}_X$. Once $\boldsymbol{p}_O$ is known, equation (9.11) and the equations which follow it may be used to find the state vector at any point in the arch. If the arch is a member of a larger structure then the method which we have just described provides a convenient systematic way of finding the equivalent loads at O and X needed in the analysis of the complete structure. These are clearly equal to $-\boldsymbol{p}_O$ and $-\boldsymbol{p}_X$.

In this particular example equation (9.15) might have been derived by the use of the compatibility method described in the previous chapter. Our present analysis corresponds, in fact, to an application of the compatibility method in which $\boldsymbol{p}_O$ is chosen as the unknown load vector $\boldsymbol{q}$, but the releases are introduced in the form of a cut at joint X. Equation (9.15) is merely the condition that the displacements associated with these releases must be zero. However, the method of transfer matrices is not, in general, just a restatement of the compatibility method in a different notation. As we shall see in the next section, it can be applied to cases where intermediate supports give a structure a high degree of redundancy, without any increase in the number of equations which have to be solved.

Although our analysis was developed for an arch with built-in ends, equation (9.14) is quite general and may be used to obtain solutions for other types of support condition. For example, a foundation settlement at X or an initial "lack of fit" due to temperature merely results in $\boldsymbol{s}_X$ having a known non-zero displacement component and equation (9.15) having a corresponding non-zero term on the left-hand side. If the arch is a plane one and the joints at O and X are pinned instead of rigid then the state vector $\boldsymbol{s}_O$ still has three known components (two zero translation components and a zero moment) and three unknown components (two forces and a rotation). Similarly the state vector $\boldsymbol{s}_X$ has three known and three unknown components, so that

$$\boldsymbol{s}_O = \begin{bmatrix} 0 \\ 0 \\ \theta_O \\ -p_{xO} \\ -p_{yO} \\ 0 \end{bmatrix}, \qquad \boldsymbol{s}_X = \begin{bmatrix} 0 \\ 0 \\ \theta_X \\ p_{xX} \\ p_{yX} \\ 0 \end{bmatrix}$$

Thus the unknowns θ_O, p_{xO}, p_{yO}, may be obtained by abstracting and solving the first, second and last scalar equations (corresponding to the zero components in $\boldsymbol{s}_X$) from the vector equation (9.14). The remaining equations may then be used to find the corresponding unknowns θ_X, p_{xX}, p_{yX}. A similar procedure may be employed in the case of a joint mounted on rollers.

9.3. A problem involving intermediate supports

We now consider the structure shown in Fig. 9.4, in which each joint has a direct connection to a rigid foundation as well as connections to adjacent joints. We assume that a known set of loads $\boldsymbol{p}_A, \ldots, \boldsymbol{p}_N$ acts at the joints $A, \ldots, N$.

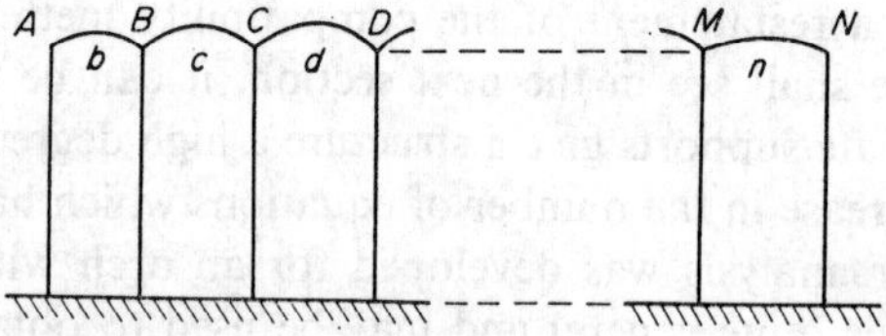

FIG. 9.4. A structure consisting of a chain of members with intermediate supports.

As before, we start our analysis at the left-hand end of the structure. Since there is no member to the left of joint A we have

$$\boldsymbol{s}_{A-} = \begin{bmatrix} \boldsymbol{d}_A \\ \boldsymbol{0} \end{bmatrix},$$

where it is now the *displacement* components of the initial state vector which are unknown. Combining equations (9.8a) and (9.9a) we obtain

$$\boldsymbol{s}_{1b}' = \boldsymbol{s}_{A+} = \begin{bmatrix} \boldsymbol{I} & \boldsymbol{0} \\ \boldsymbol{K}_A' & \boldsymbol{I} \end{bmatrix} \begin{bmatrix} \boldsymbol{d}_A \\ \boldsymbol{0} \end{bmatrix} - \begin{bmatrix} \boldsymbol{0} \\ \boldsymbol{p}_A \end{bmatrix} \quad (9.17a)$$

where $\boldsymbol{K}_A'$ is the stiffness matrix of the column supporting joint A, expressed in the global coordinate system. We write this equation as

$$\boldsymbol{s}_{1b}' = \boldsymbol{s}_{A+} = \boldsymbol{G}_A' \boldsymbol{s}_{A-} - \boldsymbol{s}_A. \quad (9.17b)$$

We now use the transfer matrix of member b to obtain

$$s_{2b}' = G_b' s_{1b}' = G_b' G_A' s_{A-} - G_b' s_A.$$

Proceeding along the structure in the same way as before we finally obtain an expression for the state vector associated with the right-hand side of joint N, which we write as s_{N+}. This expression is

$$s_{N+} = (G_N' G_n' G_M' \dots G_B' G_b' G_A') s_{A-} - (G_N' G_n' G_M' \dots G_b') s_A - \dots - s_N. \quad (9.18)$$

As before, we may add up the terms containing the known vectors s_A, ..., s_N and write equation (9.18) in the form

$$s_{N+} = \mathbf{G} s_{A-} - \mathbf{c}. \quad (9.19)$$

We now apply the boundary condition at the right-hand end of the structure, which is that the load components of s_{N+} are zero. Splitting the matrix $\mathbf{G}$ into sub-matrices $\mathbf{G}_{11}$, $\mathbf{G}_{12}$, etc., in the same way as before we write equation (9.19) in the form

$$\begin{bmatrix} d_N \\ 0 \end{bmatrix} = \begin{bmatrix} \mathbf{G}_{11} & \mathbf{G}_{12} \\ \mathbf{G}_{21} & \mathbf{G}_{22} \end{bmatrix} \begin{bmatrix} d_A \\ 0 \end{bmatrix} - \begin{bmatrix} \mathbf{c}_1 \\ \mathbf{c}_2 \end{bmatrix}$$

that is

$$d_N = \mathbf{G}_{11} d_A - \mathbf{c}_1, \quad (9.20)$$

$$0 = \mathbf{G}_{21} d_A - \mathbf{c}_2. \quad (9.21)$$

It is now equation (9.21) which has to be solved for the unknown displacement d_A. This is an "equilibrium" rather than a "compatibility" equation. Once d_A has been found, the state vectors at intermediate joints may easily be derived by working forward along the chain.

9.4. A comparison between the transfer matrix and equilibrium methods

The examples discussed in the two previous sections show that the method of transfer matrices has some of the characteristics of both the equilibrium and the compatibility methods. The final set of equations which has to be solved may be compatibility equations (as in (9.15)) or equilibrium equations (as in (9.21)). In the last example the matrix **G** was seen to be the product of both flexibility and stiffness matrices—flexibility sub-matrices appearing in the transfer matrices of the members and stiffness sub-matrices appearing in the transfer matrices of the joints.

The value of the transfer matrix method lies in the fact that the final matrix **G** is no larger than the individual transfer matrices associated with particular members or joints. It is therefore an attractive method for the structural analyst using a computer with a small working store. It must be remembered, however, that equations such as (9.15) and (9.21) merely give the unknown components of the state vector at the extreme left-hand end of the structure. To find the state vectors at intermediate points one must either repeat the formation of the partial matrix products $G_b'G_A'$, $G_B'G_b'G_A'$, etc., or record them separately during the formation of the matrix **G**.

At the end of the previous section we stated that equation (9.21) was essentially an "equilibrium" equation, since it can be written

$$\mathbf{G}_{21}\boldsymbol{d}_A = \boldsymbol{p}_{N+} - \mathbf{c}_2, \tag{9.22}$$

where $\boldsymbol{p}_{N+}$ is known to be zero since there are no members to the right of joint N. An equivalent (though not identical) equation for $\boldsymbol{d}_A$ appears during the solution of the load/displacement equations derived by the equilibrium method, and it is interesting to make a rough comparison between the amount of numerical effort required by the two methods for the example of Section 9.3.

In making this comparison we shall restrict our attention to the computation of the matrix $\mathbf{G}_{21}$ and the equivalent matrix which is obtained in the equilibrium method. These matrices are the same size as the $\boldsymbol{H}'$ and $\boldsymbol{K}'$ matrices of the individual members, i.e. 3×3 for a plane structure and 6×6 for a space structure. The number of simple numerical operations for both the inversion of an $n \times n$ matrix and the multiplication of two $n \times n$ matrices is of the order of n^3, while the addition of two such matrices requires only n^2 operations. In comparing the two methods we shall therefore ignore matrix additions and regard matrix inversion and matrix multiplication as involving roughly the same amount of computational effort. We shall term either of these processes one "basic operation" when applied to the $\boldsymbol{H}'$ or $\boldsymbol{K}'$ matrices of individual members. We shall assume that all the individual G' matrices required by the transfer matrix method and the K' matrices used in the equilibrium method are already available, and we shall not consider the numerical effort involved in their formation.

In the transfer matrix method we only need to compute the element $\mathbf{G}_{21}$ of the repeated product $\mathbf{G} = G_N' G_n' G_M' G_m' \ldots G_B' G_b' G_A'$. We may do this most efficiently by starting at the right-hand end and computing only the first column of each matrix product. Thus we begin by evaluating the first column of the product $G_b' G_A'$, and if we do not count multiplication by 0 or I this requires two basic operations. We next compute the first column of the product $G_B' G_b' G_A'$ by pre-multiplying the result of the previous step by G_B'. Again ignoring multiplication by 0 or I this requires only one basic operation. Multiplication by G_c' now requires three basic operations and multiplication by G_C' one basic operation, these numbers being repeated for each succeeding pair of member and joint transfer matrices. Thus the total number of operations required to compute $\mathbf{G}_{21}$ in this example is $12(1 + 3) + 1 + 2 = 51$.

Turning now to the equilibrium method of analysis, we may use the rules given in Section 5.2 to write the load/displacement equations of the structure in the form

$$\begin{bmatrix} p_A \\ p_B \\ p_C \\ \cdot \\ \cdot \\ \cdot \\ p_N \end{bmatrix} = \begin{bmatrix} (\Sigma K')_A & (K_{12}')_b & 0 & \cdot & \cdot & \cdot & 0 \\ (K_{21}')_b & (\Sigma K')_B & (K_{12}')_c & 0 & \cdot & \cdot & 0 \\ 0 & (K_{21}')_c & (\Sigma K')_C & (K_{12}')_b & 0 & \cdot & 0 \\ \cdot & \cdot & \cdot & \cdot & \cdot & \cdot & \cdot \\ 0 & \cdot & 0 & (K_{21}')_m & (\Sigma K')_M & (K_{12}')_n & \cdot \\ 0 & \cdot & \cdot & 0 & (K_{21}')_n & (\Sigma K')_N \end{bmatrix} \begin{bmatrix} d_A \\ d_B \\ d_C \\ \cdot \\ \cdot \\ \cdot \\ d_N \end{bmatrix} \tag{9.23}$$

where $(\Sigma K')$ indicates the sum of the direct stiffness matrices of the members meeting at a joint. These equations are symmetric and their banded form makes them very easy to solve by the elimination method described in Section 6 of the Appendix.

On this occasion it is appropriate to reverse the normal scanning sequence and start with the last equation. The first step is to eliminate the off-diagonal element $(K_{12}')_n$. We do this by adding $-(K_{12}')_n (\Sigma K')_N^{-1}$

times the last row of the matrix to the last row but one. This involves forming the product $-(K_{12}')_n(\Sigma K')_N^{-1}(K_{21}')_n$ and adding it to the sub-matrix $(\Sigma K')_M$—it will be noted that the resulting sub-matrix is still symmetric. This step requires three basic operations—an inversion and two multiplications. We do not need to compute the altered value of $(K_{12}')_n$ since the expression $(K_{12}')_n - (K_{12}')_n(\Sigma K')_N^{-1}(\Sigma K')_N$ is identically zero.

The same process may be applied to the equation associated with joint M, and so on until $(K_{12}')_b$ is eliminated from the first equation. Three basic operations are required for each matrix equation, so that the total number of operations required is $3 \times 13 = 39$—approximately 80% of the number required in the transfer matrix method. It must be admitted that this comparison is an extremely rough one. We have equated matrix inversion with matrix multiplication from the point of view of effort involved and have ignored the operations which have to be carried out on the loading terms. However, the difference is sufficient to justify the statement that the equilibrium method is at least as efficient as the method of transfer matrices.

At the beginning of this section we mentioned that the equations derived by the two methods, though equivalent, are not identical. The elimination process described above eventually reduces the first equation of (9.23) to a modified equilibrium equation for joint A,

$$K_A{}^* d_A = p_A + f(p_B, \ldots, p_N) \tag{9.24}$$

where $K_A{}^*$ is the sub-matrix which was initially $(\Sigma K')_A$ and $f(p_B, \ldots, p_N)$ represents the linear function of the applied loads which is added to p_A during the elimination process. The matrix $K_A{}^*$ is symmetric, and represents the *direct* stiffness of the complete structure as measured at joint A. If the structure is a physically sensible one then $K_A{}^*$ is likely to be well conditioned. Equation (9.22), on the other hand, is essentially an equilibrium equation for joint N, even though it contains the unknown displacement d_A. The matrix $\mathbf{G}_{21}$ is unsymmetric and represents the *cross stiffness* between joints A and N, while the vector $\mathbf{c}_2$ represents a loading system equivalent to the applied loads as far as joint N is concerned. Such an equation may not be a "good" equation for determining the displacement d_A, since if there is a very stiff support or a very flexible member at some intermediate point in the structure the equilibrium of joint N may be only very slightly affected by changes in the displacement of joint A. This point is discussed further in Section 11.4.

The method of transfer matrices is very similar to a method often used in the numerical integration of ordinary differential equations when some of the boundary conditions are specified at one end of an interval and some at the other—the so-called "two-point boundary-value problem". In dealing with such problems it is common practice to give arbitrary values to the boundary variables which are unknown at one end of the interval. The numerical integration is then carried out using these (along with the known boundary values at that end) as starting values, the calculation being repeated with revised values of the unknowns until the boundary conditions at the other end of the interval are also satisfied. Here again one is determining the values of parameters associated with a solution at one end of an interval by applying conditions at the other end. It is clear that serious numerical difficulties will arise if a solution dies away very rapidly, since whatever boundary conditions are assumed at one end the solution will have become effectively zero by the time the other end is reached.

9.5. Rigid supports and flexible joints

As stated in Section 9.1, the account of the transfer matrix method which we have given so far does not cover the case where a structure has a rigid support at some point along its length, since this corresponds to an infinite coefficient in one of the $\boldsymbol{K}'$ matrices. As mentioned in the last section, even a large finite stiffness may cause trouble by making the final matrix $\mathbf{G}$ ill-conditioned. In the same way a pin-joint connecting two members represents an infinite flexibility and gives rise to an infinite coefficient in an $\boldsymbol{F}'$ matrix. We now describe a technique which overcomes both these difficulties.

We illustrate the idea by considering the continuous beam shown in Fig. 9.5, which is built-in to a rigid support at A and simply supported at C and D. The latter supports are assumed to offer no rotational restraint, but to be infinitely stiff as far as vertical movement is concerned, while the pin at B is assumed to be incapable of transmitting a moment. The loading on the beam is taken to be equivalent to a vertical force p_{yB} at B and moments m_C, m_D at C and D. Each section of the beam has length L and

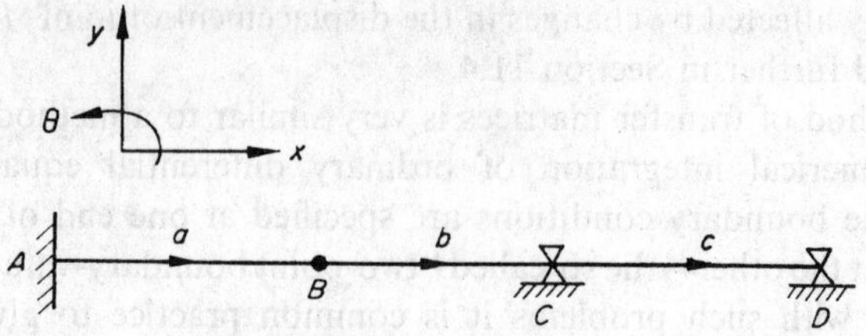

FIG. 9.5. Continuous beam with pin-joint at B and rigid supports at C and D.

flexural rigidity EI. Global and member coordinates are assumed to be identical.

In this example we ignore longitudinal forces and displacements, so that each state vector has only four components, being of the general form

$$\mathbf{s} = \begin{bmatrix} \delta_y \\ \theta \\ p_y \\ m \end{bmatrix}$$

As usual, we start the analysis at the left-hand end of the beam, where the state vector consists of two known (zero) displacement components and an unknown force and moment. Thus we have

$$\mathbf{s}_{A+} = \mathbf{s}_{1a} = \begin{bmatrix} 0 \\ 0 \\ -p_{yA} \\ -m_A \end{bmatrix}$$

where p_{yA} and m_A are the unknown reactions at the support A.

The transfer matrix of each section of beam may be obtained by dropping the rows and columns associated with axial force and displacement from the expressions given in Section 9.1. Thus we obtain

$$\mathbf{s}_{2a} = \begin{bmatrix} (\delta_{y2})_a \\ (\theta_2)_a \\ (p_{y2})_a \\ (m_2)_a \end{bmatrix} = \begin{bmatrix} 1 & L & -L^3/6EI & L^2/2EI \\ 0 & 1 & -L^2/2EI & L/EI \\ 0 & 0 & 1 & 0 \\ 0 & 0 & -L & 1 \end{bmatrix} \begin{bmatrix} 0 \\ 0 \\ -p_{yA} \\ -m_A \end{bmatrix} \quad (9.25)$$

We now have to cross the pin at B, expressing $\mathbf{s}_{1b}$ in terms of $\mathbf{s}_{2a}$. This presents no difficulty as far as vertical translation and vertical force are concerned, since we have $(\delta_{y1})_b = (\delta_{y2})_a$ and $(p_{y1})_b + (p_{y2})_a = p_{yB}$ but we have no equation which will give us $(\theta_1)_b$ from $(\theta_2)_a$. However, this lack of information about rotations is compensated for by the fact that we have a known (zero) value for the moment at B, so that $(m_2)_a = (m_1)_b = 0$, and if we put this information into (9.25) we see that it implies $Lp_{yA} - m_A = 0$, as indeed is obvious from elementary statical considerations. In other words, the unknowns p_{yA} and m_A are not linearly independent, and it is therefore unnecessary to keep them as separate variables during the remainder of the analysis. Thus we may combine the columns multiplying p_{yA} and m_A in equation (9.25) and write that equation as

$$\begin{bmatrix} (\delta_{y2})_a \\ (\theta_2)_a \\ (p_{y2})_a \\ (m_2)_a \end{bmatrix} = \begin{bmatrix} 1 & L & -L^2/3EI \\ 0 & 1 & -L/2EI \\ 0 & 0 & -1/L \\ 0 & 0 & 0 \end{bmatrix} \begin{bmatrix} 0 \\ 0 \\ m_A \end{bmatrix} \tag{9.26}$$

We have still not defined $(\theta_1)_b$. We do this simply by introducing the discontinuity in rotation $\theta_B = (\theta_1)_b - (\theta_2)_a$ as an extra unknown in equation (9.26), in place of the quantity p_{yA} which we have eliminated. Thus (9.26) becomes

$$\begin{bmatrix} (\delta_{y2})_a \\ (\theta_2)_a \\ (p_{y2})_a \\ (m_2)_a \end{bmatrix} = \begin{bmatrix} 1 & L & -L^2/3EI & 0 \\ 0 & 1 & -L/2EI & 0 \\ 0 & 0 & -1/L & 0 \\ 0 & 0 & 0 & 0 \end{bmatrix} \begin{bmatrix} 0 \\ 0 \\ m_A \\ \theta_B \end{bmatrix}$$

We can now write the state vector s_{1b} associated with the right-hand side of the hinge as

$$s_{1b} = \begin{bmatrix} (\delta_{y1})_b \\ (\theta_1)_b \\ -(p_{y1})_b \\ -(m_1)_b \end{bmatrix} = \begin{bmatrix} (\delta_{y2})_a \\ (\theta_2)_a \\ (p_{y2})_a \\ (m_2)_a \end{bmatrix} + \begin{bmatrix} 0 \\ \theta_B \\ -p_{yB} \\ 0 \end{bmatrix}$$

or as

$$s_{1b} = \begin{bmatrix} 1 & L & -L^2/3EI & 0 \\ 0 & 1 & -L/2EI & 1 \\ 0 & 0 & -1/L & 0 \\ 0 & 0 & 0 & 0 \end{bmatrix} \begin{bmatrix} 0 \\ 0 \\ m_A \\ \theta_B \end{bmatrix} - \begin{bmatrix} 0 \\ 0 \\ p_{yB} \\ 0 \end{bmatrix}$$

We now multiply this equation by the transfer matrix of member b to obtain the state vector $\boldsymbol{s}_{2b}$ associated with the left-hand side of the support C. Having done this we are faced with very much the same kind of difficulty as we encountered at B. We have continuity of rotation $(\theta_1)_c = (\theta_2)_b$ and a straightforward moment equilibrium equation $(m_2)_b + (m_1)_c = m_C$, but we have no equation giving the vertical reaction p_{yC}. We deal with this difficulty in the same way as before. We apply the condition $(\delta_{y2})_b = 0$ to obtain a relationship between m_A and θ_B, and we use this equation to eliminate one of these variables (say m_A) from the expression for $\boldsymbol{s}_{2b}$. In place of this variable we put the unknown reaction p_{yC} and then express the state vector $\boldsymbol{s}_{1c}$ associated with the right-hand side of joint C in terms of the unknowns θ_B, p_{yC} and the known applied loads at B and C. Finally we multiply this expression for $\boldsymbol{s}_{1c}$ by the transfer matrix for member c, and derive equations for the unknowns θ_B and p_{yC} by applying the conditions of zero vertical translation and known moment associated with the joint D.

9.6. Vibration problems

In Section 5.5 we saw that the application of the equilibrium method to problems of free undamped vibration leads to the problem of solving the matrix equation $\mathbf{Kd} = \omega^2\mathbf{Md}$. This is the classical eigenvalue problem and is equivalent to the problem of finding the zeros of the determinant $|\mathbf{K} - \omega^2\mathbf{M}|$.

Setting up a vibration problem in this form only involves the assembly of the matrices $\mathbf{K}$ and $\mathbf{M}$, the frequency ω appearing simply as an algebraic unknown. When transfer matrices are used the approach is somewhat different. A particular numerical value of the frequency ω is chosen and the harmonic exciting force necessary to maintain the system in vibration at this frequency is found. The natural frequencies are those values of ω for which the excitation is zero, and they are determined by a numerical root-finding process. While this procedure is somewhat inefficient compared with standard eigenvalue algorithms it is well adapted to the common practical problem of finding the response of a system over a given frequency range. The names of Holzer (torsional oscillations of shafts) and Myklestad (transverse vibrations of beams) are linked with applications of the basic idea to specific problems. However, transfer matrix notation makes it possible to present the technique in a more general form.

In the following analysis we assume that all loads and displacements vary harmonically with frequency ω, and since we are only considering undamped vibrations there are no differences of phase. Thus all state vectors have the general form $\boldsymbol{s} = \bar{\boldsymbol{s}} \sin \omega t$, and since the term $\sin \omega t$ multiplies all vectors equally we adopt the normal convention of vibration analysis and omit it from our equations. Thus state vectors in this section represent amplitudes, not instantaneous values. We consider as an example the frame shown in Fig. 9.4, which we have already considered as a static problem in Section 9.3.

The mass of the structure may be treated in several different ways. We consider first the simple "lumped mass" approximation described in Section 3.11. In this approach the distributed mass of the structure is replaced by a series of "equivalent" masses (and possibly rotational inertias) situated at the joints $A, \ldots, N$. This replacement means that the members are effectively weightless and their transfer matrices are therefore equal to the static matrices used in Section 9.3. At the joints, on the other hand, the concentrated masses produce inertia loads in place of the static external loads of the previous analysis.

If we consider a typical joint J the compatibility equation is $\boldsymbol{d}_{J+} = \boldsymbol{d}_J = \boldsymbol{d}_{J-}$, as in the static case. The equilibrium equation, on the other hand, is replaced by the equation of motion

$$\boldsymbol{p}_{J+} = \boldsymbol{p}_{J-} + \boldsymbol{K}_J'\boldsymbol{d}_J - \omega^2\boldsymbol{M}_J'\boldsymbol{d}_J$$

where $\boldsymbol{K}_J'$ is the stiffness matrix of the support and $\boldsymbol{M}_J'$ is the diagonal mass matrix associated with the joint. If the structure is a plane one then $\boldsymbol{M}_J'$ consists of the equivalent mass (in the coefficient positions associated with δ_{xJ}' and δ_{yJ}') and the equivalent rotational inertia (in the coefficient position associated with θ_J'). Combining these two equations we obtain

$$\begin{bmatrix} \boldsymbol{d}_{J+} \\ \boldsymbol{p}_{J+} \end{bmatrix} = \begin{bmatrix} \boldsymbol{I} & \boldsymbol{0} \\ \boldsymbol{K}_J' - \omega^2\boldsymbol{M}_J' & \boldsymbol{I} \end{bmatrix} \begin{bmatrix} \boldsymbol{d}_{J-} \\ \boldsymbol{p}_{J-} \end{bmatrix} \tag{9.27a}$$

which we write as

$$\boldsymbol{s}_{J+} = \boldsymbol{G}_J'(\omega)\boldsymbol{s}_{J-}. \tag{9.27b}$$

As in the earlier static analysis, we start with the state vector

$$s_{A-} = \begin{bmatrix} d_A \\ 0 \end{bmatrix},$$

where d_A is the unknown displacement of joint A. We progress along the chain in exactly the same manner as before, using the static transfer matrices defined in (9.4) for the members and the frequency-dependent transfer matrices defined in (9.27) for the joints. Eventually we obtain the state vector at the right-hand side of point N in the form

$$s_{N+} = (G_N{}'G_n{}' \ldots G_a{}'G_A{}')s_{A-} = \mathbf{G}(\omega)s_{A-}. \tag{9.28}$$

As before, we split up the matrix $\mathbf{G}(\omega)$ and write (9.28) as

$$\begin{bmatrix} d_N \\ p_{N+} \end{bmatrix} = \begin{bmatrix} \mathbf{G}_{11}(\omega) & \mathbf{G}_{12}(\omega) \\ \mathbf{G}_{21}(\omega) & \mathbf{G}_{22}(\omega) \end{bmatrix} \begin{bmatrix} d_A \\ 0 \end{bmatrix}$$

From this equation we abstract the equation $p_{N+} = \mathbf{G}_{21}(\omega)d_A$. The quantity p_{N+} represents the harmonic load which will maintain the structure in vibration at the chosen frequency, the amplitude and mode of vibration being defined by the displacement vector d_A. At the natural frequencies no excitation is required, so that $\mathbf{G}_{21}(\omega)d_A = 0$, which is equivalent to the condition $|\mathbf{G}_{21}(\omega)| = 0$. The zeros of this determinant must be found by a trial-and-error numerical method, since it is not normally practicable to construct $\mathbf{G}_{21}(\omega)$ as an *algebraic* function of ω. Solution of the equation $\mathbf{G}_{21}(\omega)d_A = 0$ for each natural frequency gives the relative magnitudes of the components of d_A (as usual in normal mode calculations, the absolute magnitude of the vector d_A may be chosen arbitrarily), and the other displacement vectors may then be found by working along the chain.

If we adopt the "consistent mass matrix" approach described in Section 3.11 we find that the transfer matrices of the members become functions of ω. In that section we showed that the dynamic load/displacement equations for a general member with distributed mass may be written in approximate form as

$$p_1 = K_{11}d_1 + K_{12}d_2 + M_{11}\ddot{d}_1 + M_{12}\ddot{d}_2,$$

$$p_2 = K_{21}d_1 + K_{22}d_2 + M_{21}\ddot{d}_1 + M_{22}\ddot{d}_2.$$

If loads and displacements vary harmonically with time then the corresponding relationship between amplitudes is

$$\left.\begin{aligned} \boldsymbol{p}_1 &= (\boldsymbol{K}_{11} - \omega^2\boldsymbol{M}_{11})\boldsymbol{d}_1 + (\boldsymbol{K}_{12} - \omega^2\boldsymbol{M}_{12})\boldsymbol{d}_2, \\ \boldsymbol{p}_2 &= (\boldsymbol{K}_{21} - \omega^2\boldsymbol{M}_{21})\boldsymbol{d}_1 + (\boldsymbol{K}_{22} - \omega^2\boldsymbol{M}_{22})\boldsymbol{d}_2. \end{aligned}\right\} \tag{9.29}$$

It is purely a matter of algebra to rearrange these equations in transfer matrix form. The procedure is similar to that carried out in Section 9.1, although since the mass matrices defined in equation (3.44) cannot be written in the form $\boldsymbol{M}_{ij} = \boldsymbol{H}_i\boldsymbol{M}\boldsymbol{H}_j^t$ the matrix products do not simplify as they did in the static case. Once the member transfer matrices have been determined as functions of ω the analysis follows exactly the same pattern as before.

We now turn to the special case of structures consisting of straight uniform members. While it is possible to use the approximate "lumped mass" or "consistent mass matrix" approaches for such structures it is just as easy to use transfer matrices which are exact solutions of the differential equations of motion. The exact transfer matrix of a straight uniform beam is actually simpler to compute than the approximate matrix derived from (9.29). Since axial and transverse vibrations are uncoupled they may be treated separately. We discuss axial vibrations first.

We consider a straight uniform beam of length L made of material with density ρ. Let $u(x,t)$ be the axial displacement of a point distant x from the origin. The displacement u satisfies the differential equation

$$\frac{\partial^2 u}{\partial t^2} = a^2 \frac{\partial^2 u}{\partial x^2}$$

where $a = \sqrt{E/\rho}$ is the longitudinal wave velocity. The harmonic solution of this equation is $u = U(x) \sin \omega t$, where $U(x) = c_1 \sin \beta x + c_2 \cos \beta x$ and $\beta = \omega/a$. The end conditions are

$$\begin{aligned} &\text{end 1:} \quad \delta_{x1} = U(0), &&\text{end 2:} \quad \delta_{x2} = U(L), \\ &\qquad -p_{x1} = EA\ U'(0), &&\qquad p_{x2} = EA\ U'(L). \end{aligned}$$

Using the conditions at end 1 we obtain the solution

$$U(x) = -(p_{x1}/EA\beta) \sin \beta x + \delta_{x1} \cos \beta x$$

and putting $x = L$ we obtain expressions for δ_{x2} and p_{x2}, which may be written in transfer matrix form as

$$\begin{bmatrix} \delta_{x2} \\ p_{x2} \end{bmatrix} = \begin{bmatrix} \cos \beta L & (\sin \beta L)/EA\beta \\ -EA\beta \sin \beta L & \cos \beta L \end{bmatrix} \begin{bmatrix} \delta_{x1} \\ -p_{x1} \end{bmatrix}$$

The transfer matrix given by this equation reduces to

$$\begin{bmatrix} 1 & L/EA \\ 0 & 1 \end{bmatrix}$$

when $\beta = 0$, as we should expect from elementary statics.

The analysis for transverse vibrations in a principal plane of bending is more complex but follows exactly the same pattern. The transverse displacement $y(x,t)$ satisfies the differential equation

$$-\frac{\partial^2 y}{\partial t^2} = a^2 \frac{\partial^4 y}{\partial x^4}$$

where $a = \sqrt{EI/\rho A}$. The harmonic solution of this equation is $y = Y(x) \sin \omega t$, where $Y(x) = c_1 \sin \beta x + c_2 \cos \beta x + c_3 \sinh \beta x + c_4 \cosh \beta x$ and $\beta^2 = \omega/a$. The end conditions are

$$\begin{aligned} \text{end 1:} \quad \delta_{y1} &= Y(0), & \text{end 2:} \quad \delta_{y2} &= Y(L), \\ \theta_1 &= Y'(0), & \theta_2 &= Y'(L), \\ -p_{y1} &= -EI\, Y'''(0), & p_{y2} &= -EI\, Y'''(L), \\ -m_1 &= EI\, Y''(0); & m_2 &= EI\, Y''(L). \end{aligned}$$

Using the conditions at end 1 we obtain the solution

$$Y(x) = \frac{1}{2}\left[\left(\frac{\theta_1}{\beta} - \frac{p_{y1}}{EI\beta^3}\right) \sin \beta x + \left(\delta_{y1} + \frac{m_1}{EI\beta^2}\right) \cos \beta x \right.$$
$$\left. + \left(\frac{\theta_1}{\beta} + \frac{p_{y1}}{EI\beta^3}\right) \sinh \beta x + \left(\delta_{y1} - \frac{m_1}{EI\beta^2}\right) \cosh \beta x \right].$$

Putting $x = L$ we obtain expressions for δ_{y2}, θ_2, p_{y2} and m_2 which may be written in transfer matrix form as

$$\begin{bmatrix} \delta_{y2} \\ \theta_2 \\ p_{y2} \\ m_2 \end{bmatrix} = \begin{bmatrix} k_0 & k_1 L & -k_3 L^3/EI & k_2 L^2/EI \\ k_3 \beta^4 L^3 & k_0 & -k_2 L^2/EI & k_1 L/EI \\ -k_1 EI\beta^4 L & -k_2 EI\beta^4 L^2 & k_0 & -k_3 \beta^4 L^3 \\ k_2 EI\beta^4 L^2 & k_3 EI\beta^4 L^3 & -k_1 L & k_0 \end{bmatrix} \begin{bmatrix} \delta_{y1} \\ \theta_1 \\ -p_{y1} \\ -m_1 \end{bmatrix} \tag{9.30}$$

where

$$k_0 = \tfrac{1}{2}(\cosh \beta L + \cos \beta L), \qquad k_1 = \frac{1}{2\beta L}(\sinh \beta L + \sin \beta L),$$

$$k_2 = \frac{1}{2\beta^2 L^2}(\cosh \beta L - \cos \beta L), \qquad k_3 = \frac{1}{2\beta^3 L^3}(\sinh \beta L - \sin \beta L).$$

It is easy to show that $k_0, k_1 \to 1$, $k_2 \to 1/2$, $k_3 \to 1/6$ as $\beta \to 0$. It follows that for $\beta = 0$ the transfer matrix in (9.30) becomes

$$\begin{bmatrix} 1 & L & -L^3/6EI & L^2/2EI \\ 0 & 1 & -L^2/2EI & L/EI \\ 0 & 0 & 1 & 0 \\ 0 & 0 & -L & 1 \end{bmatrix}$$

which agrees with (9.25).

The text by Pestel and Leckie mentioned earlier contains a comprehensive catalogue of both static and dynamic transfer matrices. In applying the method it should be remembered that the remarks in Section 9.3 about the possibility of the matrix **G** being ill-conditioned apply equally to static and dynamic problems. In cases where a rigid support or a pin-joint occurs the technique described in Section 9.5 may be applied in exactly the same way as for the static case.

CHAPTER 10

The Analysis of Non-linear Structures

Certain types of structure, particularly those formed partly from flexible cables, behave in a non-linear manner under working loads, while almost all structures have a non-linear load/displacement characteristic as they approach collapse. In this chapter we consider some of the problems which arise when we use matrix methods to investigate non-linear behaviour. Before discussing these problems, however, we review the assumptions and linearizing approximations which have been made in previous chapters.

The first of these is the assumption about the relationship between the element stress-resultant vector $\boldsymbol{r}$ and the element deformation vector $\boldsymbol{e}$. This relationship has been written as $\boldsymbol{r} = \boldsymbol{Ke}$ or $\boldsymbol{e} = \boldsymbol{Fr}$, and it has been assumed that the element stiffness matrix $\boldsymbol{K}$ (and its inverse $\boldsymbol{F}$) has constant coefficients. It is obvious that if any of the individual elements of a structure have $\boldsymbol{K}$ matrices whose coefficients depend on the element deformation then the structure will behave in a non-linear manner.

The second is the approximation which occurs in the construction of the element equilibrium equations $\boldsymbol{p}_i = \boldsymbol{H}_i\boldsymbol{r}$ expressing the nodal loads $\boldsymbol{p}_i$ in terms of the element stress-resultant vector $\boldsymbol{r}$. It will be recalled that in Chapters 3 and 4 the matrices $\boldsymbol{H}_i$ were always derived by considering the geometry of the *unloaded* element, rather than the geometry of the *deformed* element in its true equilibrium state. This procedure is justified by the fact that in most structures the deformations produced by working loads are small compared with the dimensions of the element.

A similar assumption is made when we change from element coordinates to global coordinates. We have consistently assumed that the orientation of the element coordinate axes to the global axes is independent of deformations, so that each transformation matrix $\boldsymbol{T}$ is constant and depends only on the initial undistorted geometry.

These assumptions and approximations are repeated when we derive the displacement compatibility equation for an element. This equation was obtained by considering a set of virtual nodal displacements $\boldsymbol{d}_i^*$ and the corresponding deformations $\boldsymbol{e}^*$, and arguing that the virtual work done by the nodal loads $\boldsymbol{p}_i$ must be equal to the virtual work done by the member stress-resultant $\boldsymbol{r}$. This leads to the equation $\boldsymbol{p}_i^t\boldsymbol{d}_i^* = \boldsymbol{r}^t\boldsymbol{e}^*$, and since $\boldsymbol{p}_i$ and $\boldsymbol{r}$ satisfy the equilibrium equation $\boldsymbol{p}_i = \boldsymbol{H}_i\boldsymbol{r}$ it follows that $\boldsymbol{H}_i^t\boldsymbol{d}_i^* = \boldsymbol{e}^*$. The replacement of the *virtual* displacements and deformations $\mathbf{d}_i^*$ and $\boldsymbol{e}^*$ by the *real* displacements and deformations $\boldsymbol{d}_i$ and $\boldsymbol{e}$ is only legitimate if the matrices $\boldsymbol{H}_i$ are constant during the real deformation. This is essentially the same assumption as before.

The assembly of the matrices $\mathbf{H}$ and $\mathbf{K}_m$ for a complete structure from the individual $\boldsymbol{H}$ and $\boldsymbol{K}$ matrices involves no further approximations.

If either the stiffness matrix $\mathbf{K}_m$ or the equilibrium matrix $\mathbf{H}$ is a function of the deformation of the structure it is usually necessary to employ an iterative solution procedure. A general iterative technique which can be applied to both skeletal and continuum structures is described in Section 10.3. First, however, we consider a problem of practical importance which is non-linear only in a very limited sense and whose solution is relatively simple.

10.1. Stiffness matrices for a straight uniform member with axial thrust

In the treatment of a straight uniform member in plane bending in Section 3.2 we assumed that the only effect of the axial force t was to produce axial strain. We now consider the effect which this force has on the bending moment distribution and hence on the components of the $\boldsymbol{K}_{ij}$ matrices associated with bending and shear.

We consider the member shown in Fig. 10.1. We denote the *compressive* force in the member by P. (This is to conform to the normal notation used in strut analyses.) It is clear that P is simply another symbol for p_{x1}, and that it must satisfy conditions of joint equilibrium like any other internal force when the member is part of a structure. However, while we are considering a single member it is legitimate to treat P as a known parameter, just like E, I and L.

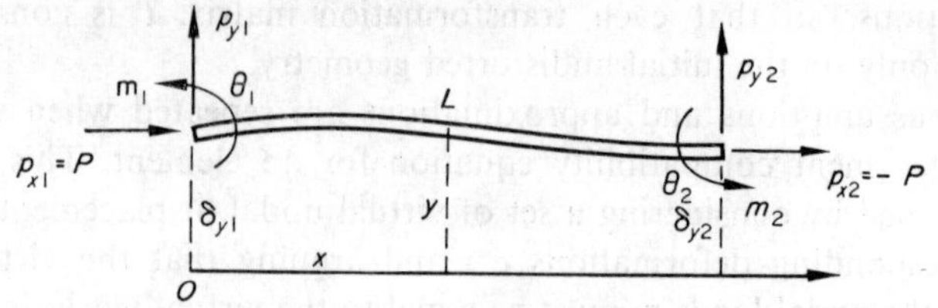

FIG. 10.1. Notation for the analysis of a beam carrying an axial thrust.

It is possible to determine the modified member K_{ij} matrices by an extension of the procedure used in Section 3.2. However, it is simpler in this instance to work from the general solution of the differential equation of bending. For the member shown in Fig. 10.1 this differential equation is

$$EI\frac{d^2y}{dx^2} = p_{y1}x - P(y - \delta_{x1}) - m_1. \tag{10.1}$$

Integrating this equation and inserting the appropriate end-conditions we obtain the modified slope-deflection equations

$$\left.\begin{aligned} m_1 &= 6\frac{EI}{L^2}\phi_2(\delta_{y1} - \delta_{y2}) + 4\frac{EI}{L}\phi_3\theta_1 + 2\frac{EI}{L}\phi_4\theta_2,\\ m_2 &= 6\frac{EI}{L^2}\phi_2(\delta_{y1} - \delta_{y2}) + 2\frac{EI}{L}\phi_4\theta_1 + 4\frac{EI}{L}\phi_3\theta_2. \end{aligned}\right\} \tag{10.2}$$

The functions ϕ_2, ϕ_3 and ϕ_4 are defined by the equations

$$\phi_2 = \tfrac{1}{3}\beta^2/(1 - \beta\cot\beta), \quad \phi_3 = \tfrac{3}{4}\phi_2 + \tfrac{1}{4}\beta\cot\beta, \quad \phi_4 = \tfrac{3}{2}\phi_2 - \tfrac{1}{2}\beta\cot\beta$$

where $\beta = (L/2)\sqrt{P/EI} = (\pi/2)\sqrt{P/P_E}$, P_E being the Euler load.

The end-shears p_{y1} and p_{y2} may be found by applying the condition of moment equilibrium

$$m_1 + m_2 + p_{y2}L + P(\delta_{y2} - \delta_{y1}) = 0. \tag{10.3}$$

It will be noticed that this equation includes the couple resulting from the difference in the lines of action of the end-forces p_{x1} and p_{x2}. Substituting for m_1 and m_2 from (10.2) gives

$$p_{y1} = -p_{y2} = \frac{12EI}{L^3}\,\phi_1(\delta_{y1} - \delta_{y2}) + \frac{6EI}{L^2}\,\phi_2(\theta_1 + \theta_2) \qquad (10.4)$$

where $\phi_1 = (\beta \cot \beta)\phi_2$. The functions $\phi_1 \ldots \phi_4$ all have value unity when $P = 0$ and are shown in Fig. 10.2.

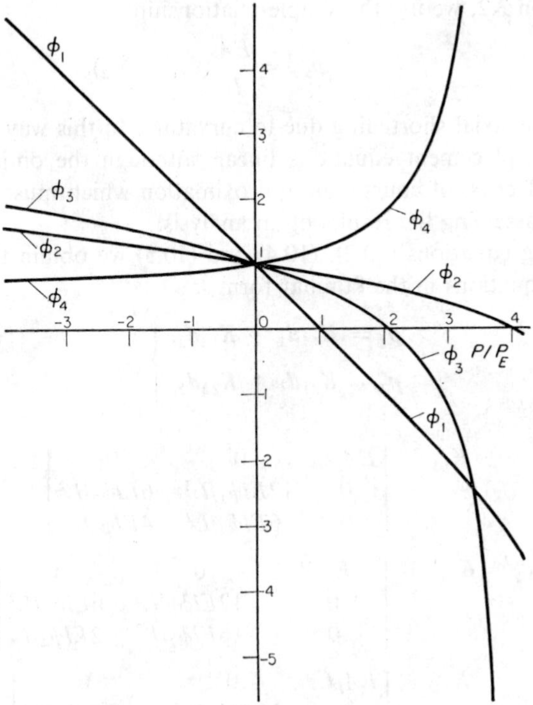

FIG. 10.2. The functions $\phi_1 \ldots \phi_4$.

The functions defined above are closely related to other stability functions defined by previous writers such as Berry and Timoshenko. Of these alternative functions the most widely used in hand calculations are the *s* and *c* functions originally introduced by Lundquist and Kroll (1944) and popularized by Merchant (Horne and Merchant, 1965). These are related to the ϕ-functions by the equations

$$s = 4\phi_3, \qquad c = \phi_4/2\phi_3.$$

Tables of these functions have been produced by Chandler and the present author (Livesley and Chandler, 1956).

In order to complete the load/displacement equations it is necessary to express the end-loads p_{x1}, p_{x2} in terms of the end-displacements δ_{x1}, δ_{x2}. As in Section 3.2, we use the simple relationship

$$p_{x1} = -p_{x2} = \frac{EA}{L}(\delta_{x1} - \delta_{x2}), \tag{10.5}$$

neglecting the axial shortening due to curvature. In this way we keep the axial load/displacement equations linear, although the omission of the curvature effect is, of course, an approximation which must be borne in mind when assessing the results of an analysis.

Combining equations (10.2), (10.4) and (10.5) we obtain the load displacement equations in the familiar form

$$\left.\begin{aligned} \mathbf{p}_1 &= \mathbf{K}_{11}\mathbf{d}_1 + \mathbf{K}_{12}\mathbf{d}_2, \\ \mathbf{p}_2 &= \mathbf{K}_{21}\mathbf{d}_1 + \mathbf{K}_{22}\mathbf{d}_2 \end{aligned}\right\} \tag{10.6}$$

where

$$\mathbf{K}_{11} = \begin{bmatrix} EA/L & 0 & 0 \\ 0 & 12EI\phi_1/L^3 & 6EI\phi_2/L^2 \\ 0 & 6EI\phi_2/L^2 & 4EI\phi_3/L \end{bmatrix}$$

$$\mathbf{K}_{12} = \mathbf{K}_{21}{}^t = \begin{bmatrix} -EA/L & 0 & 0 \\ 0 & -12EI\phi_1/L^3 & 6EI\phi_2/L^2 \\ 0 & -6EI\phi_2/L^2 & 2EI\phi_4/L \end{bmatrix}$$

$$\mathbf{K}_{22} = \begin{bmatrix} EA/L & 0 & 0 \\ 0 & 12EI\phi_1/L^3 & -6EI\phi_2/L^2 \\ 0 & -6EI\phi_2/L^2 & 4EI\phi_3/L \end{bmatrix}$$

It may be noted that equations (10.6) have the same degree of symmetry as equations (3.10). As P tends to 0 the $\mathbf{K}_{ij}$ matrices derived above tend to those defined for a simple beam in (3.11).

It is apparent that if we think of the axial force P as a known quantity then the equations (10.6) still provide a linear relationship between the end-loads and the end-displacements. They may therefore be transformed into global coordinates in the manner described in Section 3.2 and combined with similar equations to form the complete set of load-displacement equations for a structure

$$\mathbf{p} = \mathbf{K}\mathbf{d} \tag{10.7}$$

As before, the loading vector $\mathbf{p}$ may include forces and moments equivalent to loads applied at points other than the joints of the structure. However, it should be noted that the calculation of these forces and moments may be affected by the existence of axial forces in the members. For example, if a beam carrying a uniform transverse load w per unit length is subjected to an axial thrust P then it can be shown that the equivalent end-moments $wL^2/12$ are multiplied by the factor $1/\phi_2$. However, since we are assuming that the values of the axial forces in the members are known this does not materially complicate the analysis.

By solving equations (10.7) we obtain the displacements of the joints of the structure, and substituting these displacements into equations (10.6) for the various members we obtain a set of internal forces and moments. These forces and moments necessarily satisfy the equations of joint equilibrium, and to that extent are a correct analysis of the structure. It must be remembered, however, that equations (10.3) and (10.6), which the computed forces and moments also satisfy, contain the assumed values of the axial thrusts, so that the analysis is only correct (in the sense of being self-consistent) if the axial thrusts obtained as part of the computed internal forces and moments agree in value with those originally assumed when calculating the ϕ-functions for the members.

Fortunately there are many practical problems in which it is easy to make a reasonably accurate estimate of the axial forces in the members of a structure. In a triangulated rigid-jointed frame, for example, the axial forces are unlikely to differ significantly from those calculated in an analysis which treats all the joints as pinned. In rectangular frames, too, the axial forces in the stanchions may usually be estimated by treating the beams as simply-supported at their ends. In any case it is always possible to repeat the analysis using new values of the ϕ-functions based on the computed axial thrusts. This procedure usually converges adequately in two or three iterations.

10.2. The determination of elastic critical loads

The concept of the elastic critical load of a structure is essentially a generalization of the Euler buckling load of a pin-ended strut. We imagine a structure acted on by a given load system **p** such that the only internal forces are the axial thrusts P_i. We assume that these thrusts are proportional to the applied loading, so that a load system $\lambda\mathbf{p}$ produces axial thrusts λP_i. The *elastic critical load* is defined as the lowest value of λ for which the equilibrium configuration of the structure is not unique, it being possible to maintain the structure in a displaced position involving bending of the members without any additional loading, as shown in Fig. 10.3. This corresponds exactly to the normal definition of the critical

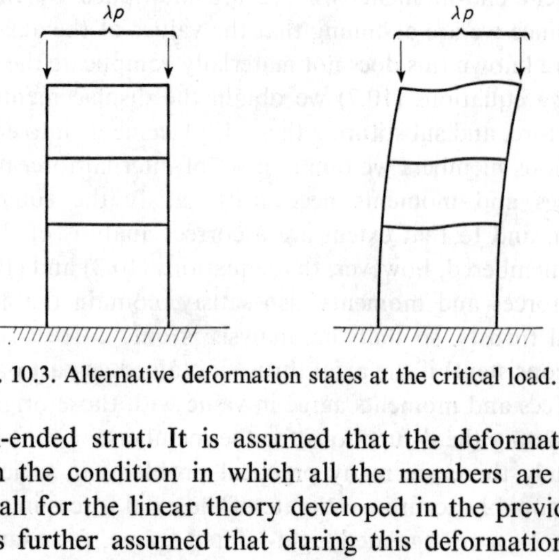

FIG. 10.3. Alternative deformation states at the critical load.

load of a pin-ended strut. It is assumed that the deformation of the structure from the condition in which all the members are straight is sufficiently small for the linear theory developed in the previous section to apply. It is further assumed that during this deformation the axial forces do not alter in value.

In general a small change in displacement $\delta\mathbf{d}$ will require the application of a small disturbing load $\delta\mathbf{p}$, these two quantities being related by equation (10.7). Since we are assuming that this additional load does not alter the values of the axial forces in the structure these forces remain equal to λP_i, so that the stiffness matrix of the structure can be regarded as a function of λ only. Thus we may write equation (10.7) in the form

$$\delta\mathbf{p} = \mathbf{K}(\lambda)\delta\mathbf{d} \tag{10.8}$$

where $\mathbf{K}(\lambda)$ is independent of $\delta\mathbf{d}$. At the critical load we have by definition $\delta\mathbf{p} = \mathbf{0}$ for some non-zero $\delta\mathbf{d}$, so that equation (10.8) becomes

$$\mathbf{K}(\lambda_{crit})\delta\mathbf{d} = \mathbf{0}. \tag{10.9}$$

It follows that the stiffness matrix $\mathbf{K}(\lambda_{crit})$ must be singular, so that λ_{crit} is the smallest positive root of the equation

$$|\mathbf{K}(\lambda)| = 0. \tag{10.10}$$

Equation (10.10) may be solved numerically by evaluating the determinant for a sequence of values of λ, the root being found by one of the standard procedures for inverse interpolation.

Alternatively an approximate solution of (10.10) may be obtained as follows. For small values of β it is easy to show that the functions $\phi_1 \ldots \phi_4$ are given approximately by

$$\phi_1 = 1 - \frac{1}{10}\frac{PL^2}{EI}, \qquad \phi_2 = 1 - \frac{1}{60}\frac{PL^2}{EI},$$

$$\phi_3 = 1 - \frac{1}{30}\frac{PL^2}{EI}, \qquad \phi_4 = 1 + \frac{1}{60}\frac{PL^2}{EI}.$$

It follows that (10.6) can be written in the approximate form

$$\left.\begin{aligned} \boldsymbol{p}_1 &= (\boldsymbol{K}_{11} - P\,\overline{\boldsymbol{K}}_{11})\boldsymbol{d}_1 + (\boldsymbol{K}_{12} - P\,\overline{\boldsymbol{K}}_{12})\boldsymbol{d}_2 \\ \boldsymbol{p}_2 &= (\boldsymbol{K}_{21} - P\,\overline{\boldsymbol{K}}_{21})\boldsymbol{d}_2 + (\boldsymbol{K}_{22} - P\,\overline{\boldsymbol{K}}_{22})\boldsymbol{d}_2 \end{aligned}\right\} \tag{10.11}$$

where $\boldsymbol{K}_{11}$, $\boldsymbol{K}_{12}$, etc., are the matrices defined in (3.11) for a beam without axial thrust and $\overline{\boldsymbol{K}}_{11}$, $\overline{\boldsymbol{K}}_{12}$, etc., are given by

$$\left.\begin{aligned} \overline{\boldsymbol{K}}_{11} &= \begin{bmatrix} 0 & 0 & 0 \\ 0 & 6/5L & 1/10 \\ 0 & 1/10 & 2L/15 \end{bmatrix} \qquad \overline{\boldsymbol{K}}_{22} = \begin{bmatrix} 0 & 0 & 0 \\ 0 & 6/5L & -1/10 \\ 0 & -1/10 & 2L/15 \end{bmatrix} \\ \overline{\boldsymbol{K}}_{12} &= \overline{\boldsymbol{K}}_{21}{}^t = \begin{bmatrix} 0 & 0 & 0 \\ 0 & -6/5L & 1/10 \\ 0 & -1/10 & -L/30 \end{bmatrix} \end{aligned}\right\} \tag{10.12}$$

These matrices have been derived by a different method by Przemieniecki (1968). He refers to them as *geometric* matrices.

Using equations (10.11) for each member the load/displacement equations for a structure may be set up in the usual way. Corresponding to (10.8) we have

$$\delta\mathbf{p} = (\mathbf{K} - \lambda\overline{\mathbf{K}})\delta\mathbf{d}$$

so that the critical load of the structure is given by the condition

$$|\mathbf{K} - \lambda\overline{\mathbf{K}}| = 0.$$

This is a linear eigenvalue problem which may be solved by any of the standard methods.

The method described above provides a rapid means of determining an approximate value of λ_{crit}. Whether the approximation is a good one depends, of course, on the accuracy with which (10.11) describes the stiffness characteristics of each member, and this in turn depends on the value of P/P_E for the member. It will be seen from Fig. 10.2 that all the ϕ-functions are approximately linear over the range $-1 \leqslant P/P_E \leqslant +1$. (The greatest absolute error in a ϕ-function produced by using the linear approximations over this range is about 0.07.)

An alternative way of determining the critical load is to investigate the effect of a constant disturbing force as λ is increased. Besides giving the value of the critical load this method also gives a picture of the behaviour of the structure as the critical condition is approached. The disturbance should be one which excites the buckling mode associated with the critical load—in most structures it is possible to make a reasonable guess at the form which this mode takes.

For example, if we consider the frame shown in Fig. 10.3 it is reasonable to assume that this frame will buckle in sidesway, as shown in the figure. An appropriate disturbance is therefore a constant horizontal load q applied to the top of the structure, as shown in Fig. 10.4a. If δ is the horizontal displacement produced by this disturbance the ratio q/δ represents the stiffness of the structure with respect to the disturbance, and as λ is increased this stiffness will diminish. Since at the critical load a displacement δ can exist without the presence of a disturbing force, it follows that at the critical load the stiffness of the structure with respect

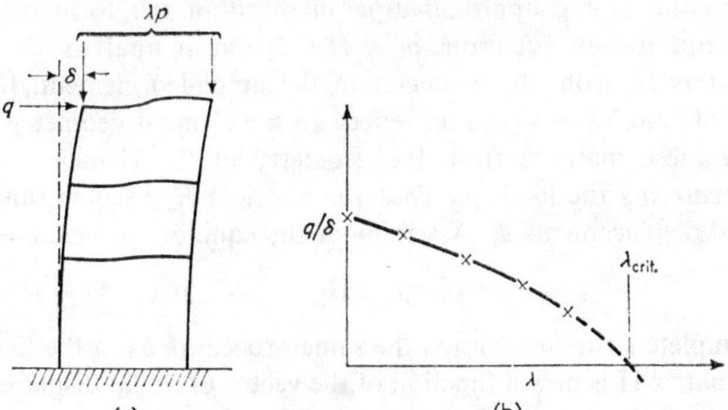

FIG. 10.4 a and b. Determination of the critical load of a structure by extrapolating the stiffness curve.

to the disturbance must be zero. Thus the critical load may be found by carrying out a series of analyses (or model experiments) at various values of λ, and extrapolating the graph of q/δ against λ until it cuts the λ-axis, as shown in Fig. 10.4b.

In applying this method it must be remembered that the use of a disturbance which does not excite the appropriate buckling mode will give an incorrect result. Thus the application of a symmetrical disturbance to the frame shown in Fig. 10.3 will give a value of λ_{crit} corresponding to the lowest symmetric buckling mode, and this may well be considerably higher than the value associated with sidesway buckling.

10.3. A general method for analysing non-linear structures

We now return to the various assumptions and linearizing approximations discussed at the start of this chapter.

The first involves the relationship between the vectors $\boldsymbol{r}$ and $\boldsymbol{e}$ for a single element. In a non-linear element the relationship $\boldsymbol{r} = \boldsymbol{Ke}$ must be replaced by the more general functional relationship $\boldsymbol{r} = f(\boldsymbol{e})$. The corresponding relationship between the vectors $\mathbf{r}$ and $\mathbf{e}$ for a complete structure is

$$\mathbf{r} = \mathrm{f}(\mathbf{e}). \tag{10.13}$$

The second is the approximation involved in the formation of the element equilibrium equations $\boldsymbol{p}_i = \boldsymbol{H}_i\boldsymbol{r}$. In linear analysis we compute the matrices $\boldsymbol{H}_i$ from the geometry of the unloaded element. If loading of the structure has a significant effect on the element geometry we must compute these matrices from the geometry of the element when it is *actually carrying* the loads $\boldsymbol{p}_i$. Thus the matrices $\boldsymbol{H}_i$ become functions of the nodal displacements $\boldsymbol{d}_i$. Assembly of the equilibrium equations

$$\mathbf{p} = \mathbf{H}\mathbf{r} \tag{10.14}$$

for a complete structure follows the same procedure as in the linear case, but the matrix $\mathbf{H}$ is now a function of the vector of nodal displacements $\mathbf{d}$. These displacements are, of course, initially unknown.

The compatibility equations for a linear element were obtained by replacing the virtual displacements and deformations in the equation $\boldsymbol{H}_i^t\boldsymbol{d}_i^* = \boldsymbol{e}^*$ by the corresponding real quantities. If the matrices $\boldsymbol{H}_i$ are functions of the nodal displacement $\boldsymbol{d}_i$ then it is only legitimate to replace $\boldsymbol{d}_i^*$ and $\boldsymbol{e}^*$ by small (strictly infinitesimal) real changes from the particular state with which the matrices $\boldsymbol{H}_i$ are associated. Thus the element compatibility equations must be written in the incremental form $\delta\boldsymbol{e} = \boldsymbol{H}_i^t\delta\boldsymbol{d}_i$. Corresponding to (10.14) we have the incremental equations

$$\delta\mathbf{e} = \mathbf{H}^t\delta\mathbf{d}. \tag{10.15}$$

Before describing a general iterative method of solving equations (10.13), (10.14) and (10.15) we illustrate its essential characteristics by considering a simple example with one degree of freedom. We consider a bar whose load-extension relation is non-linear, being of the form $p = f(\delta)$, as shown in Fig. 10.5. We wish to find the extension $\bar{\delta}$ associated with a given load $\bar{p}$, so that our problem is to solve the non-linear equation

$$p = f(\delta) = \bar{p}. \tag{10.16}$$

We solve this equation by the Newton–Raphson method, the successive steps of the solution being illustrated graphically in Fig. 10.5. In this method we start at some assumed value of displacement, here taken to be zero, and replace the non-linear bar by a bar with linear characteristics whose stiffness is the same at the assumed displacement. (By the term "stiffness" we mean here the slope of the load-displacement curve, i.e.

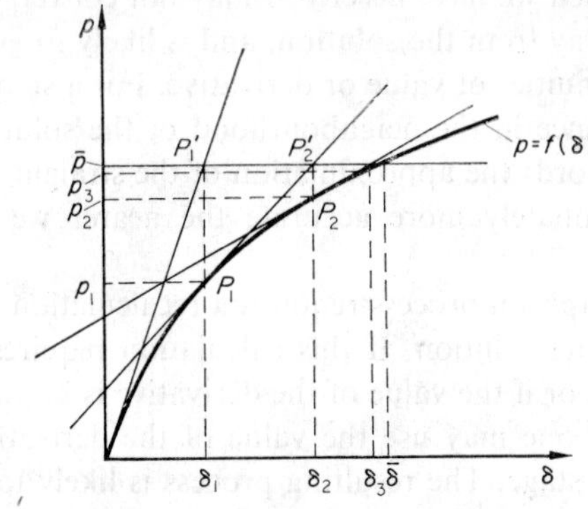

FIG. 10.5. The Newton–Raphson method for a single independent variable.

$dp/d\delta = f'(\delta)$. This is sometimes known as the *incremental stiffness*.) In graphical terms this means that we approximate to the curve by the tangent. Thus we replace (10.16) by the equation $\delta \times f'(0) = \bar{p}$, where $f'(0)$ is the derivative of the function $f(\delta)$ at the origin, and obtain an approximate solution $\delta_1 = p/f'(0)$. This corresponds to the point P_1' in Fig. 10.5. If we now substitute δ_1 into equation (10.16) we find that this displacement really corresponds to a load $p_1 = f(\delta_1)$, so that in fact we have only reached the point P_1 on the curve.

We now replace the non-linear bar by a bar with a linear load-extension law corresponding to the tangent to the curve at the point P_1, i.e. to the straight line $p = p_1 + f'(\delta_1)(\delta - \delta_1)$. Solving the linear equation $p_1 + f'(\delta_1)(\delta - \delta_1) = \bar{p}$ we obtain a second approximation

$$\delta_2 = (\bar{p} - p_1)/f'(\delta_1) + \delta_1.$$

This gives us the point P_2 in Fig. 10.4, and the process may be repeated until the difference $\bar{p} - p_n$ is less than some appropriate allowable error. Ill-conditioning may arise if the curve is nearly horizontal in the neighbourhood of the solution, since in such a case a small difference $\bar{p} - p_n$ corresponds to a much larger change in the displacement. However, if $\bar{p} - p_n$ is less in magnitude than the possible error in the specified value of $\bar{p}$ (which will only be known approximately in practice) the corresponding solution δ_n is obviously a valid solution of the problem.

The process which we have described may not converge if the starting-point is too far away from the solution, and is likely to give trouble if the curve has discontinuities of value or derivative. For a smooth curve, however, the convergence in the neighbourhood of the solution is of second order.† In other words the approximation of the straight line to the curve becomes proportionately more accurate the nearer we get to the true solution.

The Newton–Raphson process requires a recalculation of the derivative at each approximate solution. If this calculation requires a considerable amount of labour, or if the value of the derivative is known not to change a great deal, then one may use the value of the derivative calculated at the origin at each stage. The resulting process is likely to be less efficient in terms of the number of steps required to produce a solution of given accuracy, but the effort required per step is reduced. The process has only first-order convergence, and is shown graphically in Fig. 10.6. As a further alternative, one might recalculate the derivative periodically—say after every three or four steps of the iterative process. If the function $f(\delta)$ is such that it is difficult to obtain the derivative $f'(\delta)$ in algebraic form one

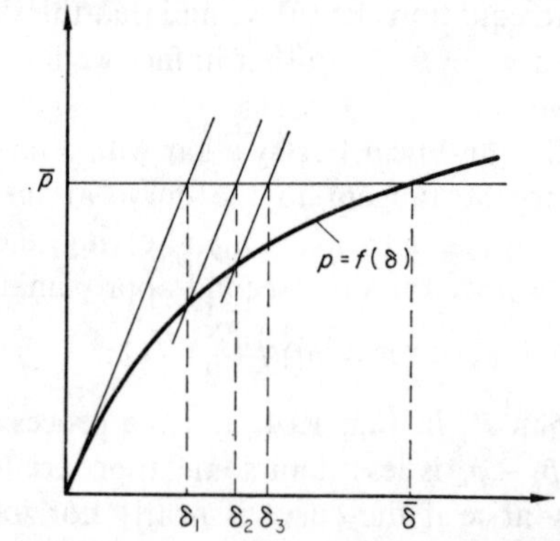

FIG. 10.6. A first-order iterative procedure.

† The order of convergence of an iterative process is defined as follows. Let $\bar{x}$ be the true solution, x_n the approximation obtained after n steps and let $\delta x_n = |\bar{x} - x_n|$. Then if $\delta x_{n+1} \approx a\,\delta x_n$, where $|a| < 1$, the convergence is first order. If $\delta x_{n+1} \approx a(\delta x_n)^2$ the convergence is second order.

may approximate to the derivative numerically, taking two points on the curve which are close together and finding the slope of the line joining them.

These ideas are easily generalized to non-linear structures with more than one degree of freedom. We begin the analysis by assuming a certain arbitrary set of deformations, and in the following account of the method we choose these deformations to be all zero. Thus we replace (10.13) by the linear approximation

$$\mathbf{r} = \mathbf{K}_{m0}\mathbf{e} \tag{10.17}$$

where $\mathbf{K}_{m0}$ represents the set of element stiffness matrices derived by ordinary linear small-displacement analysis of the individual elements. This corresponds to the tangent at the origin in Fig. 10.5.

The joint equilibrium equations associated with the assumed deformation state are

$$\mathbf{p} = \mathbf{H}_0\mathbf{r} \tag{10.18}$$

where the matrix $\mathbf{H}_0$ is based on the geometry of the undistorted structure. The displacement compatibility equations are written in the usual linear form

$$\mathbf{e} = \mathbf{H}_0{}^t\mathbf{d} \tag{10.19}$$

ignoring the change in $\mathbf{H}$ which occurs as the structure deforms. Combining (10.17), (10.18) and (10.19) we obtain the normal load-displacement equations of linear analysis $\mathbf{p} = \mathbf{H}_0\mathbf{K}_{m0}\mathbf{H}_0{}^t\mathbf{d}$, which we solve for the nodal displacements of the structure in the usual way. These displacements which we denote by $\mathbf{d}_1$, constitute our first approximation to the true solution. Substituting these displacements into (10.19) we obtain the corresponding element deformations $\mathbf{e}_1$, and the associated element stress-resultants $\mathbf{r}_1$ follow from (10.13). (It is important at this point to use (10.13) rather than the linear approximation (10.17).) We may now correct the equilibrium equations (10.18) to take account of the change in geometry, obtaining a modified set of equations

$$\mathbf{p} = \mathbf{H}_1\mathbf{r} \tag{10.20}$$

where the matrix $\mathbf{H}_1$ incorporates the displacements $\mathbf{d}_1$. In general, of course, the element stress-resultants $\mathbf{r}_1$ will not satisfy the equilibrium conditions (10.20), indicating that the solution which we have obtained requires adjustment.

The next step is to obtain a linear approximation to equation (10.13) which is valid in the neighbourhood of the deformations $\mathbf{e}_1$. We write this in the form

$$\mathbf{r} = \mathbf{r}_1 + \mathbf{K}_{m1}(\mathbf{e} - \mathbf{e}_1) \tag{10.21}$$

where $\mathbf{K}_{m1}$ represents the set of incremental member stiffness matrices associated with small changes in the deformation vector $\mathbf{e}_1$. The matrix $\mathbf{K}_{m1}$ corresponds to the slope of the load-extension curve at the point P_1 in Fig. 10.5. Corresponding to (10.20) we write the compatibility conditions in the form

$$\mathbf{e} - \mathbf{e}_1 = \mathbf{H}_1{}^t(\mathbf{d} - \mathbf{d}_1) \tag{10.22}$$

which is equivalent to (10.15). Combining equations (10.20), (10.21) and (10.22) we obtain $\mathbf{p} = \mathbf{H}_1(\mathbf{r}_1 + \mathbf{K}_{m1}\mathbf{H}_1{}^t(\mathbf{d} - \mathbf{d}_1))$, which may be written as

$$\mathbf{p} - \mathbf{H}_1\mathbf{r}_1 = \mathbf{H}_1\mathbf{K}_{m1}\mathbf{H}_1{}^t(\mathbf{d} - \mathbf{d}_1). \tag{10.23}$$

Equation (10.23) may now be solved for the vector $\mathbf{d} - \mathbf{d}_1$, which is essentially a correction to the first approximate set of nodal displacements $\mathbf{d}_1$. Adding this correction to $\mathbf{d}_1$ gives a new set of displacements $\mathbf{d}_2$, the corresponding element deformations $\mathbf{e}_2$ and element stress-resultants $\mathbf{r}_2$ being found from equations (10.22) and (10.13), respectively.

The process may be repeated until a sufficient degree of convergence has been achieved. As in the single-variable example of the Newton–Raphson process, we may measure the convergence by comparing successive displacement vectors, or alternatively we may consider successive element stress-resultant vectors. In deciding on a suitable convergence criterion it should be remembered that in a numerical application of the method there will normally be random fluctuations in successive solutions due to rounding errors in the computation, so that "exact" mathematical convergence will never be achieved.

The method as we have described it calculates successive corrections which are added to the initial approximate solution $\mathbf{d}_1$. It is possible to compute successive approximations rather than successive corrections by writing equation (10.22) in the form

$$\mathbf{p} - \mathbf{H}_1\mathbf{r} + \mathbf{H}_1\mathbf{K}_{m1}\mathbf{H}_1{}^t\mathbf{d}_1 = \mathbf{H}_1\mathbf{K}_{m1}\mathbf{H}_1{}^t\mathbf{d}.$$

The choice between the two approaches is merely a matter of computational convenience.

Although we have described this iterative method of analysing non-linear structures in terms of the equilibrium approach the compatibility method described in Chapter 8 may equally well be used for the individual linear analyses. The use of the compatibility method may be particularly advantageous if it is desired to trace the development of a collapse mechanism in a rigid-jointed frame. As far as the equilibrium approach is concerned the appearance of plastic hinges merely makes the equations more ill-conditioned, but from a compatibility viewpoint each hinge reduces the degree of redundancy, the structure being determinate when the last hinge is about to form.

10.4. The analysis of guyed masts

As a simple example of the general technique described in the previous section we consider the analysis of a transmission mast, supported by guys at a number of levels, under the action of vertical and wind loading. In this analysis we shall only consider non-linear effects due to the axial force in the mast and the load-extension characteristics of the guys—we shall ignore the effects of changes in overall geometry, although the analysis could readily be extended to cover these changes if necessary.

An example of a typical mast is shown in Fig. 10.7. As usual, we shall imagine that the distributed loads on the mast due to wind pressure and dead weight have been replaced by concentrated loads applied at the joints, the term "joints" being used here to describe the points on the mast at which the guys are attached. Although a transmission mast is essentially a space structure, in practice it is legitimate to ignore torsional effects and vertical movements of the joints due to axial strains. Thus there are four degrees of freedom at each joint—one translation and one rotation in each of the two principal planes of bending. As far as the mast sections are concerned the deformations in these two planes are quite independent, although they may be coupled as far as the complete structure is concerned by guys which do not lie in either plane. Thus the stiffness matrices

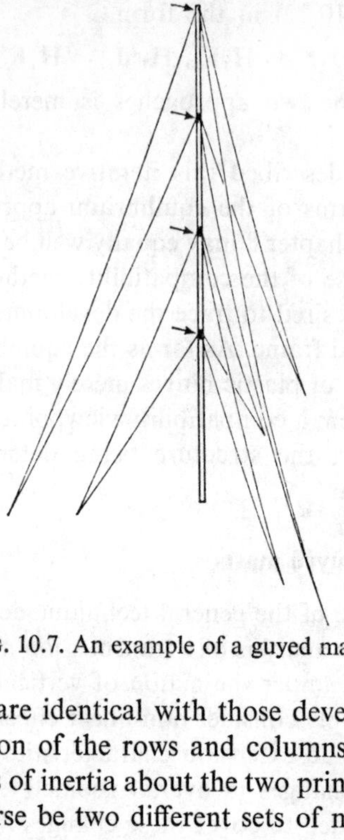

FIG. 10.7. An example of a guyed mast.

for the mast sections are identical with those developed in Section 10.1, apart from the omission of the rows and columns associated with axial strains. If the moments of inertia about the two principal axes are different then there will of course be two different sets of matrices—one for each plane.

The axial tension in any guy, which we shall denote by t, is related to the axial extension e by a curve which has the general form shown in Fig. 10.8. In the figure the condition $e = 0$ corresponds to the case where the upper end of the guy is attached to a joint whose displacement is zero. The corresponding tension t_0 is a function of the erection tension (the erection being assumed to take place in still air conditions) and the wind pressure on the guy.

In practice each guy hangs in a catenary which is comparatively flat, and we may therefore use a parabolic approximation to the catenary when deriving the force/extension equations. This results in the cubic equation

$$t^3 + t^2 (a/2t_0^2 - t_0 - EAe/L) = a/2 \qquad (10.24)$$

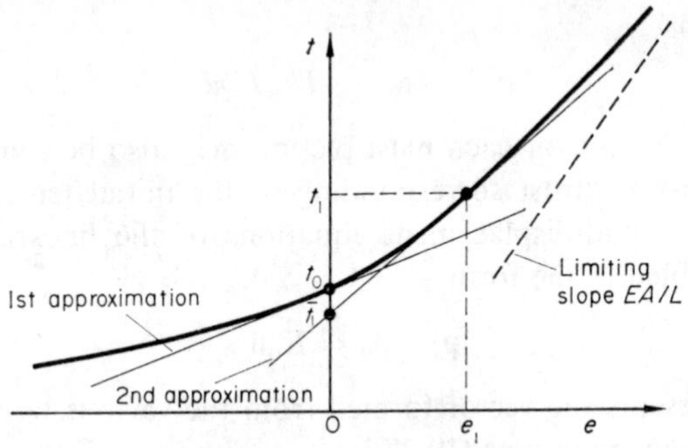

FIG. 10.8. Relation between the axial tension t and the axial extension e for a typical guy.

where a is equal to $EAW_n^2/12$, W_n being the total transverse loading on the cable due to wind and gravity forces. It will be noted that this is an implicit relation between t and e, while equation (10.16) is essentially an explicit expression for t in terms of e. However, it is easy to solve equation (10.24) numerically for any particular value of e. From that equation it is also easy to show that the slope k of the t/e curve (i.e. the stiffness of the guy at a particular load) is given by

$$k = \frac{dt}{de} = \frac{EA}{L} \frac{t^3}{t^3 + a}. \tag{10.25}$$

As might be expected, this tends to the constant value EA/L when t is large, since in this condition the guy becomes almost straight.

We begin the analysis of a mast by treating it as a linear system, using the guy stiffnesses appropriate to the assumed initial condition of zero joint displacements. Thus we imagine that each guy obeys the linear force/extension equation.

$$t = t_0 + k_0 e \tag{10.26}$$

which is transformed into system coordinates in the manner described in Section 3.4. If $\boldsymbol{p}' = \boldsymbol{T}t$ and $e = \boldsymbol{T}^t\boldsymbol{d}'$, where $\boldsymbol{p}'$ and $\boldsymbol{d}'$ are the end-load and end-displacement vectors in global coordinates,† then equation (10.26)

† In this analysis global coordinates may conveniently be defined by taking the axis of the mast as the z-axis, and making the xz and yz planes correspond to the principal planes of bending.

takes the form

$$\boldsymbol{p}' = \boldsymbol{T}t_0 + (\boldsymbol{T}k_0\boldsymbol{T}^t)\boldsymbol{d}'. \tag{10.27}$$

The stiffness matrices for each mast section may also be computed, using the values of axial thrust corresponding to the initial tensions t_0 in the guys. Thus the load-displacement equations of the linearized structure may be assembled in the form

$$\mathbf{p} - \mathbf{p}_0 = \mathbf{K}_0\mathbf{d} \tag{10.28}$$

where $\mathbf{p}_0$ represents the vector formed from the various terms $\boldsymbol{T}t_0$ which appear in the guy equations (10.27).

Equation (10.28) can now be solved to give an approximate set of displacements $\mathbf{d}_1$, and from these we can find the corresponding extensions $\mathbf{e}_1$ of the guys. Solving the cubic (10.24) for each guy with the appropriate value of e_1 we obtain the corresponding force t_1. (As mentioned in the previous section it is not advisable to use the linear approximations (10.26) at this point, since this introduces errors which are not corrected during later iterations.)

Having found the forces in the guys corresponding to the displacements $\mathbf{d}_1$, we construct for each guy a new linearized force/extension equation $t = t_1 + k_1(e - e_1)$, where the stiffness k_1 is found by substituting t_1 into equation (10.25). If we put $\bar{t}_1 = t_1 - k_1e_1$ this equation may be written as

$$t = \bar{t}_1 + k_1e \tag{10.29}$$

which is similar in form to equation (10.26). This equation for the guy force may be transformed into system coordinates in the same way as before, giving

$$\boldsymbol{p}' = \boldsymbol{T}\bar{t}_1 + (\boldsymbol{T}K_1\boldsymbol{T}^t)\boldsymbol{d}'. \tag{10.30}$$

We also use the new guy tensions t_1 to compute corrected values for the thrusts in the various sections of the mast, and calculate new stiffness matrices for these sections based on the corrected thrusts.

Assembling the joint equilibrium equations in the usual way we obtain the load/displacement equations

$$\mathbf{p} - \mathbf{p}_1 = \mathbf{K}_1\mathbf{d} \tag{10.31}$$

where $\mathbf{p}_1$ represents the vector formed from the various terms $T\bar{l}_1$ which appear in the guy equations (10.30). Solving these equations we obtain a second approximate solution $\mathbf{d}_2$, repeating the process until a satisfactory degree of convergence has been achieved. In practice the degree of non-linearity in the behaviour of a transmission mast is not great, and four or five iterations are usually sufficient to reduce the errors in the displacements and internal forces and moments to less than 1%.

It will be noticed that the process has been described here in terms of successive approximations rather than as a series of corrections to a first approximation. This makes the assembly of equations (10.28) and (10.31) identical, which is convenient from a computer programming point of view, since it means that the same section of program can be used for all iterations including the first.

If the joints of a mast are numbered in order 1, 2, . . ., N then it is apparent that joint J is connected to earth by guys and to joints $J+1$ and $J-1$ by adjacent mast sections. The linearized structure is therefore one which can conveniently be solved by the method of transfer matrices described in Chapter 9.

10.5. Non-linear problems associated with plastic collapse analysis

In previous sections of this chapter we considered purely elastic problems of non-linear behaviour. The iterative procedure which we used was essentially an extension of the linear equilibrium method described in Chapter 5. The linear theory of rigid-plastic collapse presented in Chapter 7 may be extended in a similar manner.

In Section 7.3 the rigid-plastic collapse load λ_c is defined as the maximum value of λ for which a solution of the equilibrium equations

$$\lambda \mathbf{p} = \mathbf{H}\mathbf{r} \tag{10.32}$$

exists satisfying the linear constraints

$$-\mathbf{r}^L \leqslant \mathbf{r} \leqslant \mathbf{r}^U. \tag{10.33}$$

Section 7.7 describes an algorithm for finding λ_c and the associated vector $\mathbf{r}$ from these equations. Since the structure is assumed to be completely rigid until λ_c is reached, the equilibrium equations (10.32) are those associated with the initial undistorted structure, as in ordinary linear elastic analysis.

It follows that the collapse mechanism associated with λ_c should strictly be regarded as an *infinitesimal* real displacement† from the initial state. The vectors $\delta\mathbf{e}$ and $\delta\mathbf{d}$ which define this displacement satisfy the compatibility equations

$$\delta\mathbf{e} = \mathbf{H}^t\delta\mathbf{d} \tag{10.34}$$

where **H**, as stated above, is computed from the *undistorted* geometry of the structure. If a real finite displacement occurs then the geometry changes, so that the equilibrium equations (10.32) are no longer strictly true. However, if **H** is altered so that it relates to the *deformed* geometry then a new solution of (10.32) and (10.33) may be carried out, giving the value of λ_c associated with the deformed state. By repetition of this procedure a complete picture may be obtained of the way in which λ_c varies as the structure collapses. In some cases the collapse mechanism may change as the deformation increases.

A flow diagram for a computer program to carry out this calculation is shown in Fig. 10.9. In this program the changes in the geometry of the

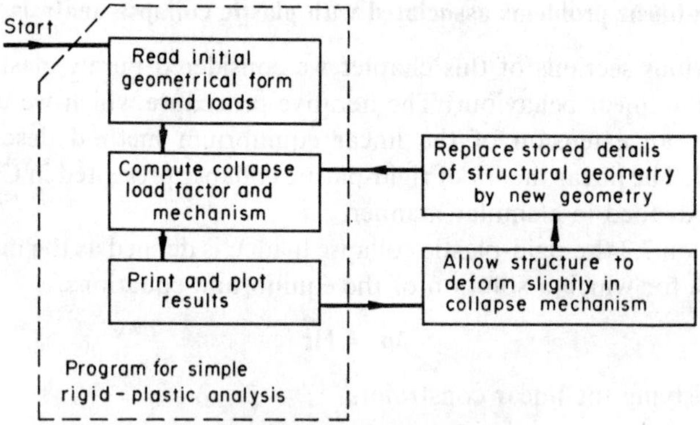

FIG. 10.9. A flow diagram for computing the effect of geometry changes on the plastic collapse load.

† Alternatively one may imagine a "virtual mechanism" of arbitrary magnitude, but this is effectively the same thing, since a virtual mechanism is based on the linearized compatibility equations associated with the undistorted structure.

structure are calculated from (10.34). Since the increments of deformation and displacement are finite the changes are not strictly compatible, but experiments with such a program indicate that a maximum plastic hinge rotation increment of 2° does not introduce significant cumulative errors in a sequence of analyses with final plastic hinge rotations of the order of 30°.

An example of the way in which λ_c varies with deformation is shown in Fig. 10.10. The vertical loads at the tops of the stanchions are an important feature of this example. The initial beam mechanism with its internal hinge causes the span of the beam to shorten, with the result that the right-hand stanchion becomes inclined to the vertical. This in turn causes the

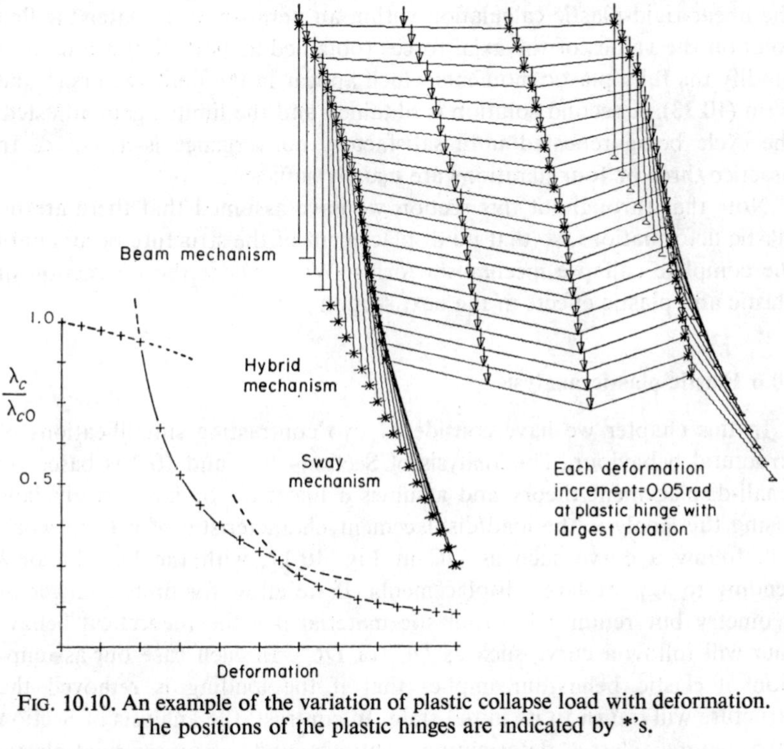

FIG. 10.10. An example of the variation of plastic collapse load with deformation. The positions of the plastic hinges are indicated by *'s.

vertical load at the top of that stanchion to induce a thrust in the beam, with the result that the structure finally collapses in a side-sway mode.

This example shows how changes in geometry can affect the collapse load, particularly when the members of a frame carry significant axial forces. Such forces also affect the initial value of λ_c, since an axial force in a member reduces the value of its fully plastic moment of resistance. For a rectangular section the value of m_p decreases in a parabolic fashion as the axial thrust P increases, while for practical rolled-steel sections the variation is more complex, being a function of the shape of the cross-section. For details of the variation of m_p with axial force the reader is referred to a standard text on plastic theory, such as Neal (1963).

This aspect of the affect of axial forces may also be analysed by setting the linear rigid-plastic calculation within an iterative loop. After the first solution the values of the axial forces (obtained as part of $\mathbf{r}$) are used to modify the fully plastic moments which appear in the limit vectors $\mathbf{r}^L$ and $\mathbf{r}^U$ in (10.33). A second solution is obtained and the limits again adjusted, the cycle being repeated until satisfactory convergence is achieved. In practice three or four iterations are usually sufficient.

Note that throughout this section we have assumed that there are no elastic deformations, so that no displacement of the structure occurs until the complete collapse mechanism forms. We consider the interaction of elastic and plastic effects in the next section.

10.6 Elastic-plastic analysis

In this chapter we have considered two contrasting simplifications of structural behaviour. The analysis of Sections 10.1 and 10.2 is based on small-displacement theory and assumes a linear-elastic stress/strain law. Using this analysis the load/displacement characteristic of a framework will follow a curve such as OC in Fig. 10.11, with the load factor λ tending to λ_{crit} at large displacements. If we allow for finite changes in geometry but retain a linear-elastic material law the theoretical behaviour will follow a curve such as OC' or OC''. In each case our assumption of elastic behaviour implies that if the loading is removed the structure will return to its initial state. In contrast, the analysis of Section 10.5 ignores elastic deformations entirely and assumes rigid-plastic

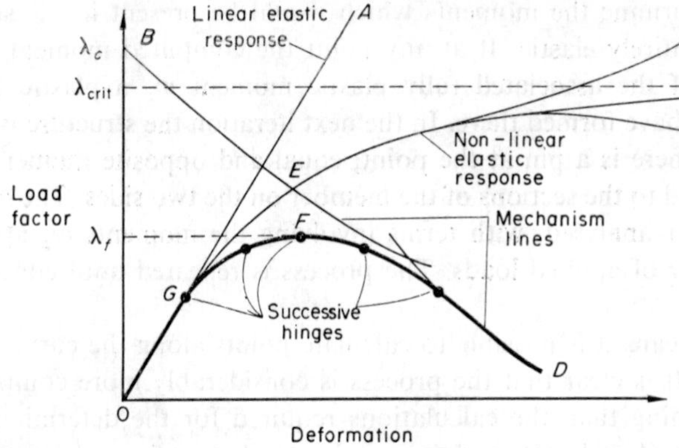

FIG. 10.11. Various approximations to the behaviour of a structure under increasing load.

material. For a structure formed from such material the load/displacement characteristic will be a curve such as *OBD*.

If all the plastic hinges in the collapse mechanism were to form simultaneously one would expect the behaviour of a structure made from ductile material to follow the curve *OED*. In reality, of course, the plastic hinges appear sequentially as the loads are increased, and the experimental curve is more likely to ressemble *OFD*. The maximum load which the structure can carry occurs at the point *F*. In view of the difficulty of calculating λ_f various authors have suggested semi-empirical methods of estimating it from the more easily determined parameters λ_c and λ_{crit}. One well-known formula, due to Merchant, is

$$\frac{1}{\lambda_f} = \frac{1}{\lambda_c} + \frac{1}{\lambda_{crit}} \tag{10.35}$$

which has the merit of simplicity, although it is sometimes conservative and sometimes unsafe. For further information on this topic the reader is referred to the text by Horne and Merchant (1965).

The iterative method described in Section 10.3 may be used to analyse a framework which is partially plastic. In the case of a framework with straight uniform members the analysis presented in Section 10.1 may be used to determine the moments which would be present in the structure if it were entirely elastic. If at any point the computed moment exceeds the value of the associated fully plastic moment m_p a plastic hinge is assumed to have formed there. In the next iteration the structure is treated as though there is a pin at the point, equal and opposite moments $\pm m_p$ being applied to the sections of the member on the two sides. The modified structure is reanalysed, with terms involving the moments m_p appearing in the vector of applied loads. The process is repeated until convergence is achieved.

By this means it is possible to calculate points along the curve *OGF* in Fig. 10.11. It is clear that the process is considerably more complex and time-consuming than the calculations required for the determination of the curves *OC* and *BD*, and it is unlikely that such analysis will ever become part of normal design practice. Our main reason for describing it in this section is that it introduces two important general problems of non-linear analysis which also arise in the solution of elastic-plastic continuum problems.

The first problem is associated with continuity. In the transmission-mast analysis described in the previous section we dealt with structural elements whose stiffness varied with deformation in a continuous manner. The assumption that a plastic hinge suddenly appears in a member when the moment at some point reaches m_p gives rise to a discontinuity in stiffness (which does not actually exist in the real structure) which may have a disastrous effect on the convergence of the iterative process. The calculation may in certain circumstances oscillate indefinitely, while in others it may converge to an incorrect solution. This difficulty may be avoided, at the cost of added complication in the calculation of the member stiffness matrices, by making the stiffness coefficients vary in a more continuous way, and by giving the plastic hinge some positive rotational stiffness—such stiffness exists when a real member becomes plastic due to the presence of strain-hardening. In fact one may argue that the convergence troubles which have occurred in certain computer programs for elastic-plastic analysis have been due to approximations which were initially introduced to "simplify" the calculation.

The second problem is somewhat more fundamental and is associated with the non-conservative nature of partially-plastic material. The iterative procedure described in Section 10.3 arrives at the true solution of a non-linear problem by means of a series of approximate linear solutions. These solutions do not normally have any particular physical significance—they are merely successive approximations in a numerical process. Since in general these approximate solutions do not follow the growth of deformation in the real structure as the loads are applied, the process cannot be used to analyse a non-conservative structure in which the solution depends not only on the loading but also on the loading history.

This means that we encounter difficulties if we use the method to analyse a non-linear structure or continuum in which a stress reversal occurs in material which has been deformed plastically. The difficulty may be illustrated by considering a structure which contains a bar whose elastic-plastic load/extension curve is shown in Fig. 10.12. Imagine that in two successive iterations we obtain two values of extension δ_n and δ_{n+1}, where $\delta_{n+1} < \delta_n$, as shown in the figure. This may correspond to a true unloading of the bar, in which case we ought to follow the unloading curve XX', using the slope of this line at X in the iteration which produces δ_{n+1}. On the other hand, the apparent reduction in displacement may be merely a chance incident in the iteration process, without there being any unloading of the real bar at all. If this is so we clearly ought to follow the unloading curve XX'', and use the slope of *this* line at X in the iteration process.

To solve history-dependent problems of this sort it is necessary to trace

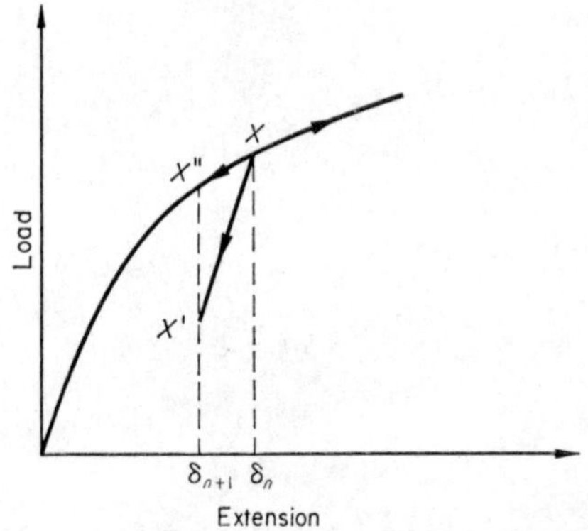

FIG. 10.12. The effect of unloading in the plastic region.

the true sequence of deformation states as the actual loading program is carried out. This requires a *sequence* of non-linear analyses, a small increment $\delta\mathbf{p}$ being applied before each analysis. After the iteration has converged a check must be made to verify that the loading and unloading stiffnesses which have been used are consistent with the signs of the computed increments in the deformations.

This procedure is still open to the objection that it does not follow the actual changes in deformation as the increment $\delta\mathbf{p}$ is applied. In fact it assumes that the behaviour up to loading $\mathbf{p}$ depends on how the loads are applied, but that this dependence may be ignored when considering the change in loading $\delta\mathbf{p}$. However, this assumption seems to be a reasonable compromise in practice and it is difficult to see how it can be avoided in any method which involves the use of finite increments of loading.

CHAPTER 11

Problems of Size and Accuracy

In this book we have developed a number of systematic methods of structural analysis suitable for use with computers. Programs based on these methods can be used to analyse structures of any size and complexity, subject only to limits imposed by the size and accuracy of the computer used. In this chapter we discuss these limits and indicate ways in which they can be extended.

If we look at the way in which computers are used in structural engineering practice we find that many design engineers who do not regard themselves as computer specialists periodically write short programs of, say, 50 to 100 statements in a high-level algebraic language such as Fortran or Algol. These are usually special-purpose programs associated with relatively simple structural problems. When we look at more complex problems we find that programming becomes a specialist activity, with engineers dividing into "program writers" and "program users". Finally when we come to the large general-purpose packages such as ICES, ASKA and GENESYS we reach the point where program design and construction becomes a specialist team activity involving computer scientists, numerical analysts and professional programmers. This chapter is intended for the program user rather than the program writer.

The equation-solving routines in large structural analysis packages are highly complex, and the practical engineer will generally be content to assume that they are logically correct, accurate and computationally efficient. However, anyone using such a package for the analysis of a large structure with, say, 1000 joints or so will do well to know something of the way large sets of linear equations are handled on a computer. How we define "large" depends, of course, on the size of computer which is available—essentially we are concerned here with structures which give rise to matrices too big to fit into the store of the computer without some form of coefficient packing. A general account of some of the techniques in current use is given in Section 11.1.

A structure which is too large to be analysed as a single system may be tackled by splitting it into two or more sub-structures. This approach was essential in the early days of automatic computers, when storage restrictions were much more severe than they are today. It is still a useful technique and is described in Section 11.2.

Modern computers are sufficiently reliable for "mistakes" of the sort made by human beings to be rare. However, the professional engineer remains responsible for the results of any calculation he produces, whether it is done by hand or by computer program. He will naturally ask, "How can I check that the results produced by the computer are correct?" In Section 11.3 we discuss suitable procedures for checking matrix calculations.

In practice, of course, all numerical work involves loss of precision due to rounding errors. By a "correct" answer we mean one where rounding errors are acceptably small. Since rounding errors are cumulative they tend to be most troublesome in the analysis of large structures with several thousand joints. However, ill-conditioning can cause computational difficulties in much smaller problems, and in Section 11.4 we consider some of the situations in which ill-conditioning may arise.

In this chapter we only consider problems of linear analysis. However, since non-linear problems are normally solved by a series of linearized approximations the ideas developed here are also relevant to the material presented in Chapter 10.

11.1. The solution of large sets of load/displacement equations

The problems associated with the solution of the load/displacement equations of complex structures are best introduced by means of specific examples. We begin this section by considering the three-dimensional frame shown in Fig. 11.1.

If we apply the general rules developed in Section 5.3 we find that the stiffness matrix of this structure has the form shown in Fig. 11.2. In this

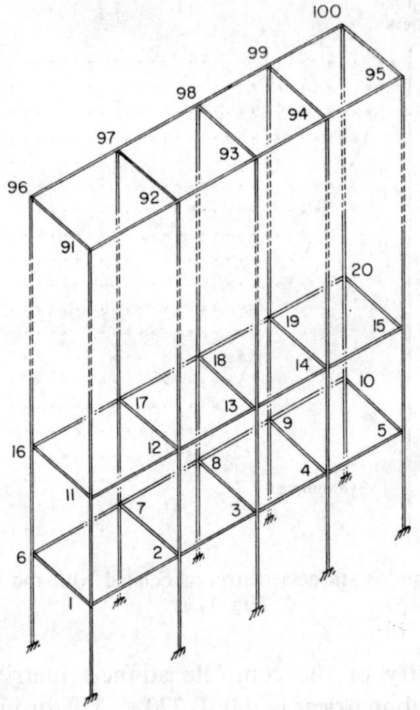

FIG. 11.1. A three-dimensional frame with 100 joints.

figure an asterisk indicates a non-zero 6×6 sub-matrix and a dot a zero sub-matrix. Since the stiffness matrix is symmetric, only the part above and including the leading diagonal has been shown. It will be recalled from Section 5.3 that the leading diagonal sub-matrix in row i corresponds to the direct stiffness of joint i, while a non-zero sub-matrix in row i, column j indicates the presence of a member connecting joints i and j. There is thus one leading diagonal sub-matrix for each joint and two symmetrically placed off-diagonal sub-matrices for each member (apart from members connected to a rigid foundation, which contribute no off-diagonal terms). It follows that for the frame shown in Fig. 11.1 there are $100 + 2 \times 220 = 540$ non-zero sub-matrices. Since the total number of sub-matrices is 10,000 the proportion which are non-zero is slightly over 5%. If we make

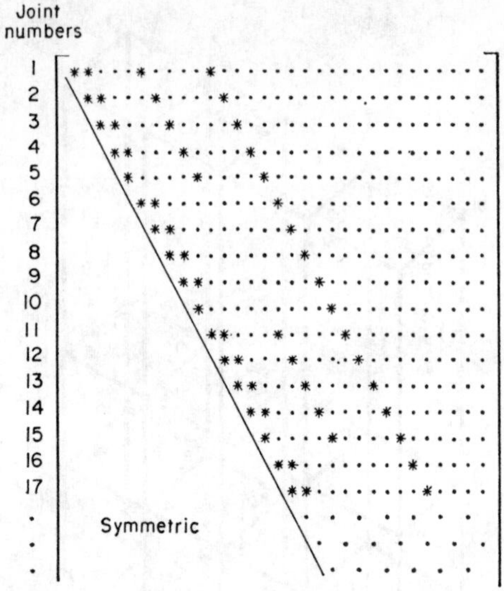

FIG. 11.2. The banded stiffness matrix associated with the frame shown in Fig. 11.1.

use of the symmetry of the complete stiffness matrix the number of distinct non-zero sub-matrices is $100 + 220 = 320$, or just over 3% of the total. For a frame with M members and N joints the proportion of non-zero coefficients in the stiffness matrix is $(2M + N)/N^2$. Since M is roughly proportional to N for structures of a given type the proportion of non-zero coefficients varies roughly as $1/N$. A similar argument may be applied to the stiffness matrices associated with large finite element problems.

This quality of "sparseness" is present in virtually every large stiffness matrix. The matrix in Fig. 11.2 also possesses another characteristic—that of having its non-zero sub-matrices arranged in bands parallel to the leading diagonal. We define the *bandwidth* of a matrix to be the maximum number of sub-matrices in a row, from the sub-matrix on the leading diagonal to the one in an outermost band, inclusive. In our example the bandwidth is 11 sub-matrices (i.e. 66 scalar coefficients).

The banded pattern shown in Fig. 11.2 is a consequence of the regular nature of the structure in Fig. 11.1 and the regular system of joint numbering. In general the bandwidth is equal to 1 + the maximum difference between the pairs of joint numbers at the ends of the members of a framework (or the set of node numbers associated with the element vertices in a finite element problem). It is clear that for a given structure there will be a particular arrangement of numbering which produces the smallest bandwidth. In the case of a regular structure such as the one we have chosen for our example this arrangement is usually obvious.

The ideal computer routine for solving large sets of load/displacement equations is one which minimizes both computer storage requirements and computing time, though in practice the two minima are unlikely to be attained simultaneously. The design of such a routine is largely a matter of taking advantage of the qualities of sparseness and bandedness discussed above. As an introduction to the ideas involved we consider the organization of an elimination procedure for symmetric banded systems of equations. (An account of the numerical operations which make up this procedure is given in Section 6 of the Appendix.)

Since the elimination method alters the coefficient matrix, it may seem at first sight as though space must be provided for all coefficients, since those initially zero may become non-zero during the calculation. However, it is easy to see that while most of the zeros inside the outermost bands will be altered, the zeros outside these bands will not be affected. We illustrate the practical consequences of this fact by again considering the example in Fig. 11.2.

We begin by considering the question of storage. If the pivots in the elimination process are chosen sequentially from the leading diagonal then after equation i has been used as a pivot the remaining $n-i$ equations in $n-i$ variables are still symmetric about the leading diagonal. Since zeros outside the outermost bands are never altered it follows that it is only necessary to store, for each row, the coefficients from the leading diagonal to one of the outermost bands—usually the one in the upper triangle. In the example chosen this implies that only 66 coefficients need to be stored for each scalar equation. This gives us a total storage requirement of approximately 40,000 scalar coefficients. While this is larger than the figure of 320 non-zero sub-matrices (11,000 scalar coefficients) mentioned earlier it is still only 11% of the storage required for the complete matrix.

We now consider the way in which this storage is used during the elimination process. At each stage we have a pivotal equation with 66 coefficients. Because of the banded nature of the matrix this equation only has to modify, at most, the next 65 equations—the remaining equations already have a zero in the appropriate column. Thus during each pivotting operation we are only concerned with $66^2 \approx 8000$ coefficients out of the total 40,000. This is illustrated in Fig. 11.3.

Imagine now that we have a computer with a working store† sufficient to hold 8000 numbers and with auxiliary storage in the form of magnetic tape or discs. The calculation may be arranged in the following way. The first 66 equations are transferred from the auxiliary store to the working store of the computer, and the first equation is used to modify the remaining 65 in the usual way. When this modification has been completed the first equation is transferred back to the auxiliary store and is replaced by the 67th equation. Since the second equation and all the other equations on which it has to operate are now in the working store the process can be repeated using the second equation without any further transfers. When the operations based on the second equation are complete it too is returned to the auxiliary store and replaced by the 68th equation, and so on. In this scheme each equation is brought from the auxiliary store to the working store and takes its place at the end of a queue. While in the working store

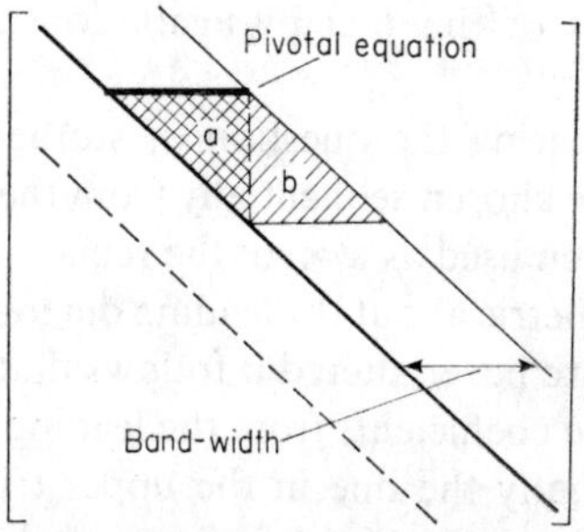

FIG. 11.3. Region of a banded matrix modified by a given pivotal equation. Region (a)—coefficients which are actually altered. Region (b)—unaltered coefficients in modified equations.

† By the phrase "working store" we mean that part of the computer where arithmetic operations are carried out on numbers. "High-speed store" and "core store" are alternative names.

it is modified by all the equations which have higher positions in the queue, until eventually it reaches the position where it is used as the pivotal equation. After it in its turn has been used to modify the equations which follow, it is returned to the auxiliary store. During the elimination each equation is only transferred between the auxiliary and the working stores once in each direction. The final back-substitution involves transferring the modified equations to the working store in the reverse order.

In the procedure outlined above the most important quantity is the bandwidth b, rather than the total number of equations n. The amount of working storage required is b^2, while the number of arithmetic operations required is approximately $nb^2/2$ (in contrast to the figure of $n^3/3$ for a matrix without zeros). For a given bandwidth the amount of effort required is proportional to the number of equations—if the frame shown in Fig. 11.1 were extended upwards for another ten floors the computing time (and the auxiliary storage required) would be doubled, but the size of working store needed would not be affected.

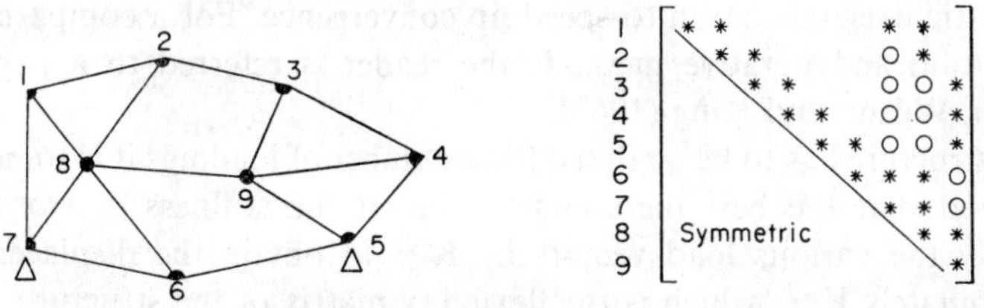

FIG. 11.4. An example of a structure with a sparse stiffness matrix. An asterisk indicates a non-zero sub-matrix. A circle indicates a zero sub-matrix which becomes non-zero during the elimination.

The elimination method may also be applied to sparse matrices which are not banded. Figure 11.4 shows a simple example. Once again it will be seen that the method only alters certain of the zero sub-matrices. General routines for handling sparse systems of equations involve considerably more complex programming logic than the sequential scheme outlined above and we shall not discuss details here. One of the best-known schemes is the "frontal" method due to Irons (1970). For a general account of such methods the reader is referred to a book by Tewarson (1973).

In the banded matrix procedure both the storage and computing time depend on the bandwidth, which is a function of the way the nodes of the structure are numbered. In the more general case of a sparse matrix without noticeable bands the number of zero sub-matrices which become non-zero during the elimination also depends on the disposition of the off-diagonal sub-matrices and hence on the node numbering. The reader may like to experiment with different numbering schemes for the frame shown in Fig. 11.4. Various automatic methods of node numbering have been developed which attempt to produce an optimum or near-optimum arrangement of the matrix (i.e. minimum bandwidth for a banded matrix routine, minimum total storage for a sparse matrix routine).

Although elimination methods are the ones most commonly employed in general structural analysis programs the Gauss–Seidal and other iterative procedures are also used. These have the advantage that the coefficient matrix is not altered during the calculation, so that only the non-zero coefficients need to be stored. While these procedures use minimal computer storage they tend to require rather more computing time than elimination methods. They are normally coupled with techniques such as systematic over-relaxation to speed up convergence. For a comparison of elimination and iterative methods the reader is referred to a paper by Clough, Wilson and King (1963).

If a structure has to be analysed for a number of loadings it is sometimes suggested that the best method is to invert the stiffness matrix $\mathbf{K}$ and multiply the various load vectors by $\mathbf{K}^{-1}$ to obtain the displacements. Unfortunately $\mathbf{K}^{-1}$, which is the flexibility matrix of the structure, is unlikely to have many zeros, even when $\mathbf{K}$ is heavily banded. A better procedure is to carry out an elimination solution of the equations, processing all the loading vectors simultaneously. This means that although each loading requires a separate back-substitution the reduction to triangular form (which consumes the bulk of the time) need only be done once.

11.2. The use of sub-structures

The techniques described in the previous section are quite general and may be applied to any sparse and banded system of equations. Another concept which is useful in the analysis of complex structures is that of the sub-structure.

Once again we introduce the idea by means of an example, shown in Fig. 11.5a. We imagine that this structure is too large to be analysed as a single unit with the computer available. We therefore regard it as two simpler structures, connected at the points M and N. (While the division between the two structures is to some extent arbitrary the approach is most effective if the division is made at a point where the number of connecting joints is as few as possible.) These two structures we term sub-structures.

We now adopt a procedure similar to that used in Chapter 8, inserting releases at the points connecting the two systems and applying self-equilibrating load-pairs at these points. By doing this we reduce the structure to the two simpler structures shown in Fig. 11.5b, which can be considered quite separately. We begin the analysis by writing the load/displacement equations of substructure 1 in the form

$$\begin{bmatrix} \boldsymbol{p}_A \\ \cdot \\ \cdot \\ \boldsymbol{p}_L \\ \boldsymbol{q}_M \\ \boldsymbol{q}_N \end{bmatrix} = \begin{bmatrix} \\ \\ & \mathbf{K}_1 & \\ \\ \\ \end{bmatrix} \begin{bmatrix} \boldsymbol{d}_A \\ \cdot \\ \cdot \\ \boldsymbol{d}_L \\ \boldsymbol{d}_M \\ \boldsymbol{d}_N \end{bmatrix} \tag{11.1}$$

Since $\boldsymbol{q}_M$ and $\boldsymbol{q}_N$ are unknown we cannot solve these equations completely

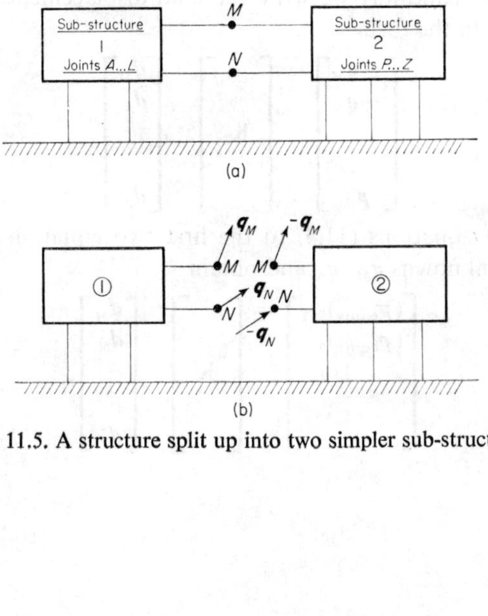

FIG. 11.5. A structure split up into two simpler sub-structures.

but we can commence the elimination process in the usual way, subtracting suitable multiples of the first equation from the others, etc., until we have eliminated all coefficients of the joint displacements $\boldsymbol{d}_A, \ldots, \boldsymbol{d}_L$ from the equilibrium equations for joints M and N. Thus we reduce the last two equations of (11.1) to

$$\begin{bmatrix} \boldsymbol{q}_M + f_M(\boldsymbol{p}_A, \ldots, \boldsymbol{p}_L) \\ \\ \boldsymbol{q}_N + f_N(\boldsymbol{p}_A, \ldots, \boldsymbol{p}_L) \end{bmatrix} = \begin{bmatrix} & & \\ & \mathbf{K}_1{}^* & \\ & & \end{bmatrix} \begin{bmatrix} \boldsymbol{d}_M \\ \\ \boldsymbol{d}_N \end{bmatrix} \tag{11.2}$$

Here $\mathbf{K}_1{}^*$ indicates the altered coefficients and f_M, f_N the linear combinations of the known applied loads at joints $A, \ldots, L$ which are added to the unknowns $\boldsymbol{q}_M$ and $\boldsymbol{q}_N$ during the elimination process. We write equations (11.2) in the form

$$\begin{bmatrix} \boldsymbol{q}_M \\ \\ \boldsymbol{q}_N \end{bmatrix} = \begin{bmatrix} & & \\ & \mathbf{K}_1{}^* & \\ & & \end{bmatrix} \begin{bmatrix} \boldsymbol{d}_M \\ \\ \boldsymbol{d}_N \end{bmatrix} - \begin{bmatrix} (\boldsymbol{p}_{\text{equiv}})_{M1} \\ \\ (\boldsymbol{p}_{\text{equiv}})_{N1} \end{bmatrix} \tag{11.3}$$

These equations have exactly the same form as equations (3.40). The reduced stiffness matrix $\mathbf{K}_1{}^*$ is symmetric and represents the stiffness of sub-structure 1 as viewed from joints M and N, while the loads $(\boldsymbol{p}_{\text{equiv}})_{M1}$ and $(\boldsymbol{p}_{\text{equiv}})_{N1}$ are the loads which the structure would exert on the joints M and N if these joints were prevented from displacing.

In exactly the same way we write the load/displacement equations for sub-structure 2 in the form

$$\begin{bmatrix} -\boldsymbol{q}_M \\ -\boldsymbol{q}_N \\ \cdot \\ \cdot \\ \boldsymbol{p}_Z \end{bmatrix} = \begin{bmatrix} & & \\ & \mathbf{K}_2 & \\ & & \end{bmatrix} \begin{bmatrix} \boldsymbol{d}_M \\ \boldsymbol{d}_N \\ \cdot \\ \cdot \\ \boldsymbol{d}_Z \end{bmatrix} \tag{11.4}$$

If we now add equations (11.3) to the first two equations of (11.4) we eliminate the unknowns $\boldsymbol{q}_M, \boldsymbol{q}_N$ and obtain

$$\begin{bmatrix} (\boldsymbol{p}_{\text{equiv}})_{M1} \\ (\boldsymbol{p}_{\text{equiv}})_{N1} \\ \cdot \\ \cdot \\ \boldsymbol{p}_Z \end{bmatrix} = \begin{bmatrix} & & \\ & \mathbf{K} & \\ & & \end{bmatrix} \begin{bmatrix} \boldsymbol{d}_M \\ \boldsymbol{d}_N \\ \cdot \\ \cdot \\ \boldsymbol{d}_Z \end{bmatrix} \tag{11.5}$$

where **K** is the matrix formed by adding $\mathbf{K}_1{}^*$ to the appropriate elements of $\mathbf{K}_2$. Since the loads in equations (11.5) are all known, these equations may be solved for the displacements of sub-structure 2 without difficulty. Once the displacements of joints *M* and *N* are known we may drop the last two equilibrium equations in (11.1) and solve the remaining equations for the displacements of sub-structure 1.

An alternative approach is to take the equilibrium equations for sub-structure 2 and eliminate the displacements of joints *P*, . . ., *Z*, in the same way as we did for sub-structure 1. This gives

$$\begin{bmatrix} -q_M + (p_{\text{equiv}})_{M2} \\ -q_N + (p_{\text{equiv}})_{N2} \end{bmatrix} = \begin{bmatrix} \mathbf{K}_2{}^* \end{bmatrix} \begin{bmatrix} d_M \\ d_N \end{bmatrix} \tag{11.6}$$

where $(p_{\text{equiv}})_{M2}$ and $(p_{\text{equiv}})_{N2}$ are the equivalent loads at joints *M* and *N* due to the loads on sub-structure 2. Equation (11.6) defines the behaviour of sub-structure 2 as viewed from the joints *M* and *N*. If we add equations (11.3) and (11.6) our analysis is effectively the same as before. If on the other hand we write equations (11.3) and (11.6) in the form

$$\begin{bmatrix} d_M \\ d_N \end{bmatrix} = \begin{bmatrix} \mathbf{K}_1{}^{*-1} \end{bmatrix} \begin{bmatrix} q_M + (p_{\text{equiv}})_{M1} \\ q_N + (p_{\text{equiv}})_{N1} \end{bmatrix}$$

$$\begin{bmatrix} d_M \\ d_N \end{bmatrix} = \begin{bmatrix} \mathbf{K}_2{}^{*-1} \end{bmatrix} \begin{bmatrix} -q_M + (p_{\text{equiv}})_{M2} \\ -q_N + (p_{\text{equiv}})_{N2} \end{bmatrix}$$

we may obtain the unknowns q_M, q_N by equating the displacements of the joints *M* and *N* in the two sub-structures, writing the equations in the form

$$\begin{bmatrix} \mathbf{K}_1{}^{*-1} + \mathbf{K}_2{}^{*-1} \end{bmatrix} \begin{bmatrix} q_M \\ q_N \end{bmatrix} = - \begin{bmatrix} \mathbf{K}_1{}^{*-1} \end{bmatrix} \begin{bmatrix} (p_{\text{equiv}})_{M1} \\ (p_{\text{equiv}})_{N1} \end{bmatrix} + \begin{bmatrix} \mathbf{K}_2{}^{*-1} \end{bmatrix} \begin{bmatrix} (p_{\text{equiv}})_{M2} \\ (p_{\text{equiv}})_{N2} \end{bmatrix}$$

This equation may be solved for q_M and q_N, after which we may obtain the detailed solution for each sub-structure by solving equations (11.1) and (11.4).

The first approach described above (i.e. the addition of equations (11.3) and (11.4)) is merely a computational device to allow the elimination solution of the equilibrium equations of the complete structure to be done in two parts. The second approach, in which we apply compatibility conditions at joints M and N to determine the unknown internal loads $\boldsymbol{q}_M$ and $\boldsymbol{q}_N$, is an application of the compatibility method described in Chapter 8. There is, however, one important difference between the analysis developed here and the analysis given in that chapter. In the previous account of the compatibility method we introduced sufficient releases into a structure to make it determinate. It was therefore possible to determine the flexibility matrix of the resulting structure merely from the equations of joint equilibrium and the individual flexibilities of the members. In the analysis presented in this section both sub-structures may well be hyperstatic, so that it is necessary to find their flexibility matrices by inverting their stiffness matrices.

The idea of a sub-structure is nothing more than a generalization of the concept of a structural element, and all we are really doing is regarding the sub-structure as our basic "building brick". Following the procedure we used in the case of simple elements, we focus our attention on the displacements of the nodes which connect the various sub-structures together. Thus we replace loads acting at internal points of the sub-structures by "equivalent" loads at the connecting nodes. The equations (11.3) are identical to the load-displacement equations of a simple structural element expressed in "stiffness" form, and may be used to build up the load/displacement equations of a more complicated structure in exactly the same way. For example, consider the arrangement of three sub-structures shown in Fig. 11.6. Following the notation used in dealing with simple elements, the suffixes 1 and 2 denote the points at which sub-structures are connected to other sub-structures. The number of degrees of freedom at the various connections may be different, although the number on each side of a particular connection must necessarily be the same. If we take the equilibrium equations for each sub-structure and eliminate all the "internal" displacements and loads we obtain equations which we can write in the form

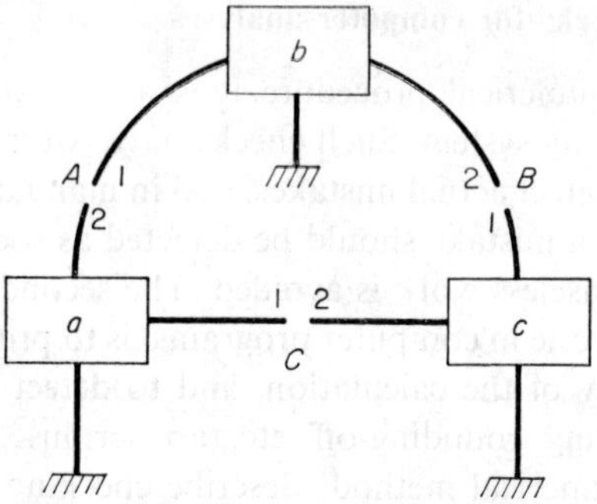

FIG. 11.6. A structure formed from three sub-structures.

$$\boldsymbol{p}_{1a} = (\boldsymbol{K}_{11})_a \boldsymbol{d}_{1a} + (\boldsymbol{K}_{12})_a \boldsymbol{d}_{2a} - (\boldsymbol{p}_{\text{equiv}})_{1a},$$
$$\boldsymbol{p}_{2a} = (\boldsymbol{K}_{21})_a \boldsymbol{d}_{1a} + (\boldsymbol{K}_{22})_a \boldsymbol{d}_{2a} - (\boldsymbol{p}_{\text{equiv}})_{2a},$$

etc., the equations being similar for the other sub-structures. (Although primes have been omitted from these equations they should be regarded as being in a common global coordinate system.) If we now apply compatibility and equilibrium conditions at the connections A, B and C we obtain the equations

$$\begin{bmatrix} (\boldsymbol{p}_{\text{equiv}})_A \\ (\boldsymbol{p}_{\text{equiv}})_B \\ (\boldsymbol{p}_{\text{equiv}})_C \end{bmatrix} = \begin{bmatrix} (\boldsymbol{K}_{22})_a + (\boldsymbol{K}_{11})_b & (\boldsymbol{K}_{12})_b & (\boldsymbol{K}_{21})_a \\ (\boldsymbol{K}_{21})_b & (\boldsymbol{K}_{22})_b + (\boldsymbol{K}_{11})_c & (\boldsymbol{K}_{12})_c \\ (\boldsymbol{K}_{12})_a & (\boldsymbol{K}_{21})_c & (\boldsymbol{K}_{11})_a + (\boldsymbol{K}_{22})_c \end{bmatrix} \begin{bmatrix} \boldsymbol{d}_A \\ \boldsymbol{d}_B \\ \boldsymbol{d}_C \end{bmatrix}$$

where $(\boldsymbol{p}_{\text{equiv}})_A = (\boldsymbol{p}_{\text{equiv}})_{2a} + (\boldsymbol{p}_{\text{equiv}})_{1b}$, etc. Once these equations have been solved for the connecting displacements $\boldsymbol{d}_A$, $\boldsymbol{d}_B$ and $\boldsymbol{d}_C$ we may return to a detailed analysis of each individual sub-structure if necessary.

This replacement of an assembly of members by an "equivalent" system, in which only the joints which connect the assembly to the remainder of the structure are considered, is particularly useful when a structure contains a prefabricated member made up from a large number of smaller parts—a braced girder is an obvious example. The idea is not really a new one—it is simply an application of the basic finite element approach on a larger scale. The reader familiar with electric network analysis will realize that what we have been discussing is nothing more than a structural equivalent of Thevenin's theorem.

11.3. Numerical checks for computer analyses

In designing a numerical procedure it is always desirable to include some kind of checking system. Such checks have two main functions. The first function is to detect actual mistakes, and in manual work in particular it is important that a mistake should be detected as soon as possible after it is made, so that useless work is avoided. The second function, which is the most important one in computer programs, is to provide an estimate of the overall accuracy of the calculation, and to detect cases where errors due to ill-conditioning, rounding-off, etc., are serious.

Textbooks on numerical methods describe checking systems which are appropriate for matrix calculations when these are carried out by hand. In programming structural problems for a computer, however, it is common practice not to include any programmed checks on the individual matrix operations, but to provide an overall check on the final results. If the analysis is based on the equilibrium method then the final step involves the calculation of the internal loads from the element stiffness matrices and the computed displacements. The simplest way of checking these loads is to verify that they are in equilibrium with the external loads applied at the nodes. In essence this is simply a substitution of the computed nodal displacements $\mathbf{d}$ into the original equations $\mathbf{p} = \mathbf{Kd}$ to verify that they do in fact satisfy the equations.

In practice, of course, the conditions of equilibrium will never be satisfied exactly. Instead of the true displacements $\mathbf{d}$ we inevitably calculate some approximation to this set, which we may call $\mathbf{d}_1$, and when this is multiplied by $\mathbf{K}$ we obtain a load vector $\mathbf{p}_1$ which differs from $\mathbf{p}$ by some amount $\delta\mathbf{p} = \mathbf{p} - \mathbf{p}_1$. The vector $\delta\mathbf{p}$ is termed the *residual vector*. Having calculated $\delta\mathbf{p}$ we may if necessary improve the accuracy† of the solution by solving the equation

$$\delta\mathbf{p} = \mathbf{K}\,\delta\mathbf{d} \tag{11.7}$$

thus obtaining a correction $\delta\mathbf{d}$ which is added to $\mathbf{d}_1$. Of course this correction itself is only approximate, but if necessary the process may be re-

† This process merely corrects errors due to rounding-off during the solution of the equations. It does not, of course, correct rounding errors made during the calculation of $\mathbf{K}$ itself.

peated until sufficient accuracy has been achieved. The approach is very similar to the method of analysing non-linear structures described in Section 10.3.

It will be noted that the simplest check to apply is a check on the degree to which the internal loads satisfy the nodal equilibrium equations, and not a check on the accuracy of the displacements. It is true that we may obtain an estimate of the error in the displacements by solving equation (11.7), but this will involve just as much work as the solution of the original equations. It is also true that if the matrix $\mathbf{K}$ is ill-conditioned then the error vector $\delta\mathbf{d}$ may be considerably larger than the residual vector $\delta\mathbf{p}$. However, we must not overlook the fact that in any real problem of structural analysis the load vector $\mathbf{p}$ is only known approximately. If from the computed displacements $\mathbf{d}_1$ we calculate a load vector $\mathbf{p}_1$, the first question we should ask is whether we really know $\mathbf{p}$ so precisely that the difference $\delta\mathbf{p}$ is significant. If $\delta\mathbf{p}$ is less than the inevitable uncertainty in our knowledge of the applied loads any attempts to improve the "accuracy" of our solution are bound to be useless. Looking at the problem from a slightly different point of view, we may say that our solution $\mathbf{d}_1$ and the internal loads derived from it are exact for a system of applied loads $\mathbf{p}_1$. Thus we have an exact solution, not of the loading case $\mathbf{p}$ which we set out to analyse, but of a loading case $\mathbf{p}_1$ which differs from the original by a certain amount. If the difference is insignificant then the solution we have found is clearly an acceptable one.

The same type of overall check may be applied in the compatibility method. Having calculated an approximate set of member stress-resultants $\mathbf{r}_1$ the obvious thing to do is to see whether these really do produce continuity at the releases. Thus we evaluate $\mathbf{B}^t\mathbf{F}_m\mathbf{r}_1$ and see whether it is zero. As in the case of the equilibrium method, we shall find that due to rounding errors during the calculation the result will not be quite zero. Before any attempts to improve the "accuracy" of the calculation are made it is necessary to consider whether the "error" is meaningful. The error here is an error in the satisfying of compatibility conditions at the releases, just as before it was an error in satisfying equilibrium conditions at the nodes. We must accordingly ask whether we know that during the erection of the structure the compatibility conditions were satisfied to this degree of accuracy. In other words we say that our solution is "exact" for a structure in which there is a certain amount of initial lack of fit, and we consider whether or not this amount of misalignment is likely to arise in practice.

Although the checks we have described have been thought of as overall checks they are really only a test that the equilibrium equations have been solved or the flexibility matrix inverted correctly. In the equilibrium method, for example, each member $\boldsymbol{K}'$ matrix is used twice during the calculation—first in the formation of the stiffness matrix of the structure **K** and afterwards in the calculation of the element nodal forces. A computing error in the calculation of a $\boldsymbol{K}'$ matrix will lead to the final solution being incorrect, but if the faulty $\boldsymbol{K}'$ matrix is used consistently the resulting internal loads will satisfy the erroneous equilibrium equations, so that the check we have described will not detect the error. In fact we shall have obtained a correct analysis of the wrong structure. In modern computers such errors are extremely rare—certainly a great deal rarer than the occasions when the same effect is produced by a human error in the preparation of the computer input data.

Finally it must be stated that while an equilibrium check is a valuable aid to confidence in the results produced by a computer, in practice the degree of accuracy provided by an automatic computer is usually considerably more than that required in normal structural calculations. A persistent failure in a checking process indicates a genuine numerical difficulty, and implies the existence of something very unusual in the structure being analysed—or, of course, an error in the data.

11.4. Ill-conditioning in structural analysis

In analysing a structure by the equilibrium method we have to solve the set of equations

$$\mathbf{p} = \mathbf{K}\mathbf{d} \qquad (11.8)$$

and it is natural to consider in what circumstances, if any, there is likely to be computational difficulty due to ill-conditioning of the matrix **K**. Similarly in the compatibility method it is necessary to invert the matrix $\mathbf{B}^t\mathbf{F}_m\mathbf{B}$, and it is desirable to choose a release system which makes this matrix as well conditioned as possible. We consider the equilibrium method first.

To illustrate both the cause and the effect of ill-conditioning in the equilibrium method we consider the simple one-dimensional system shown in Fig. 11.7, in which three springs of stiffnesses k, K and k support two

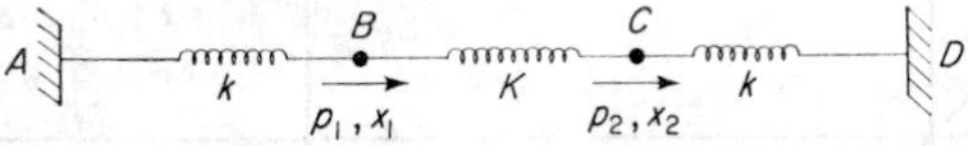

FIG. 11.7. A simple one-dimensional system illustrating ill-conditioning.

points B and C at which known loads p_1, p_2 are applied. In this example the equilibrium equations are

$$p_1 = (k + K)\, x_1 - Kx_2,$$

$$p_2 = -Kx_1 + (k + K)x_2,$$

which may be written

$$\begin{bmatrix} p_1 \\ p_2 \end{bmatrix} = \begin{bmatrix} k + K & -K \\ -K & k + K \end{bmatrix} \begin{bmatrix} x_1 \\ x_2 \end{bmatrix} \tag{11.9}$$

We see that in this case the vector

$$x = \begin{bmatrix} x_1 \\ x_2 \end{bmatrix}$$

is specified by its scalar products with the two vectors $a_1{}^t = [k + K - K]$, $a_2{}^t = [-K \;\; k + K]$. Whether these are "good" axes for defining a general two-dimensional vector depends on the relative magnitudes of the stiffnesses K and k. The directions of the vectors a_1, a_2 and the lines corresponding to the scalar equations defined by (11.9) are shown in Fig. 11.8 for the two cases $k/K = 0.1$ and $k/K = 10$.

It will be seen that when k is small relative to K the vector x is ill-defined, in the sense that small changes in the loads or the stiffness coefficients may produce large changes in x_1 and x_2. To solve equations (11.9) by the elimination method we multiply the first equation by $K/(k + K)$ and add it to the second. This gives

$$\begin{bmatrix} p_1 \\ p_2 + Kp_1/(k + K) \end{bmatrix} = \begin{bmatrix} k + K & -K \\ 0 & (k + K) - K^2/(k + K) \end{bmatrix} \begin{bmatrix} x_1 \\ x_2 \end{bmatrix}$$

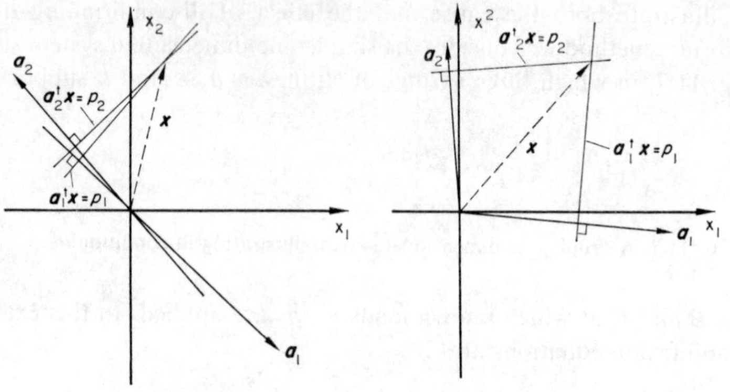

FIG. 11.8 (a). An ill-conditioned system: $k/K = 0.1$.
(b) A well-conditioned system: $k/K = 10$.

and the last equation, which gives x_2, may be written

$$p_2 + Kp_1/(k + K) = \left\{\frac{(k + K)^2 - K^2}{k + K}\right\} x_2.$$

If we consider the evaluation of the coefficient of x_2 in this equation as an algebraic process then there is, of course, no difficulty, but if the coefficients $k + K$ and K are only available in numerical form then we have to find the difference of two squares, which will be very nearly equal if $k \ll K$. Thus the coefficient of x_2 will have fewer significant figures than the original coefficients $k + K$ and K, and this inaccuracy will in general lead to both x_1 and x_2 being in error.

It is apparent that in this simple example ill-conditioning arises when the stiffness K of the spring which connects the two points B and C is large compared with the stiffness of the springs which connect these points separately to the fixed supports A and D, so that the displacements x_1 and x_2 are physically very closely linked together. In fact we may argue that the real cause of the ill-conditioning is not that the equilibrium equations are a "bad" way of defining the vector

$$\begin{bmatrix} x_1 \\ x_2 \end{bmatrix}$$

but that this vector is a "bad" way of defining the displaced form of the system when K is large. For example, if we define the displacements of the points B and C by means of the *extensions* of the springs AB and BC then the equations giving these extensions are quite well conditioned.

The same phenomenon occurs in the analysis of rigid-jointed frames by the equilibrium method, particularly in frames of unbraced form. If we consider the simple portal frame shown in Fig. 11.9a then a solution by the

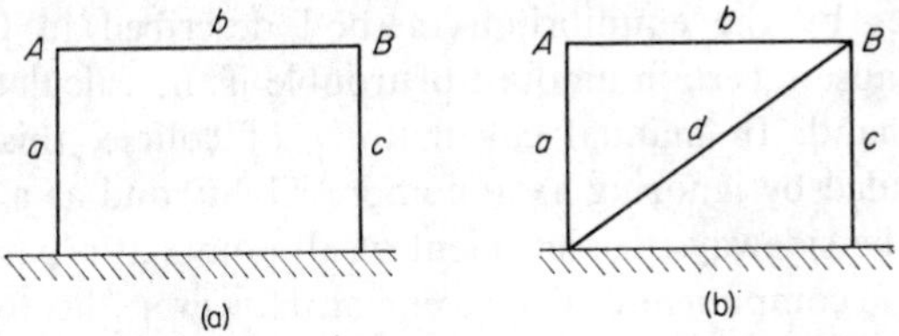

FIG. 11.9. (a) A frame with ill-conditioned sway equations. (b) A frame with well-conditioned sway equations.

equilibrium method involves defining the displacement of the structure by means of the two displacement vectors

$$\boldsymbol{d}_A = \begin{bmatrix} \delta_{xA} \\ \delta_{yA} \\ \theta_A \end{bmatrix} \qquad \boldsymbol{d}_B = \begin{bmatrix} \delta_{xB} \\ \delta_{yB} \\ \theta_B \end{bmatrix}$$

If we construct the equilibrium equations in the manner described in Chapter 5 and extract the scalar equations concerned with horizontal equilibrium we find that these have the form

$$\left.\begin{aligned} p_{xA} &= \{(EA/L)_b + (12EI/L^3)_a\}\delta_{xA} - (EA/L)_b\,\delta_{xB} \\ p_{xB} &= -(EA/L)_b\delta_{xA} + \{(EA/L)_b + (12EI/L^3)_c\}\delta_{xB} \end{aligned}\right\} \qquad (11.10)$$

In most practical frames the direct compressive stiffness terms EA/L are considerably greater than the shear stiffness terms $12EI/L^3$, so that equations (11.10) are ill-conditioned in exactly the same way as equations (11.9).

This type of ill-conditioning only appears in frames which are unbraced. If we add a diagonal bracing d to the frame shown in Fig. 11.9a, so that we obtain the frame shown in Fig. 11.9b, it will be found that the equations are quite well behaved. This is because the diagonal bracing gives joint B a direct stiffness with respect to horizontal displacements which is comparable in magnitude to the axial stiffness of the member b connecting A and B. It should be noted that the ill-conditioning depends on the stiffness matrix $\mathbf{K}$ and not on the loading vector $\mathbf{p}$. Thus it is a function of the structure, and the choice of variables used to define its deformation, rather than the particular load system applied.

The degree of ill-conditioning which appears in the solution of rigid-jointed frames by the equilibrium method described in Chapter 5 is sufficient to cause a certain amount of trouble if the calculations have to be done by hand. In manual calculations, of course, this difficulty is normally avoided by ignoring axial compressibility and as a consequence representing the sideways displacement of all points at a particular floor-level by a single component of displacement. However, this to some extent destroys the systematic nature of the method, which is its chief attraction from the point of view of an automatic computer.

Most automatic computers store numbers to a precision of about 10 decimal places, and experience shows that this is sufficient to avoid serious trouble due to ill-conditioning when using the equilibrium method for the analysis of practical frameworks, whether they are braced or not. Some experiments carried out by the author suggest that in an unbraced frame formed from ordinary rolled-steel sections the axial stiffnesses can be increased by a factor of about 10^3 before the effects of ill-conditioning become serious.† These comments are based on the assumption that the equilibrium equations are solved by the elimination method—the ill-conditioning is certainly sufficiently severe to make the simple Gauss–Seidel iteration process converge extremely slowly when applied to unbraced rigid-jointed frames.

Similar arguments may be applied to the matrix $\mathbf{B}^t\mathbf{F}_m\mathbf{B}$, which has to be inverted when the compatibility method is used. In Section 8.3 we showed that this matrix, which expresses the displacements $\mathbf{u}$ at the releases in terms of the release load-pairs $\mathbf{q}$, has the general form

† These experiments were carried out on simple frames with about twelve joints. In structures with several hundred joints round-off errors can cause trouble during the solution process, even with well-conditioned equations, so that double-length arithmetic may be necessary.

$$\begin{bmatrix} F_{11} & F_{12} & . & F_{1n} \\ F_{21} & F_{22} & . & F_{2n} \\ . & . & . & . \\ F_{n1} & F_{n2} & . & F_{nn} \end{bmatrix}$$

where the leading diagonal elements F_{11}, F_{22}, etc., represent the direct flexibilities at the various releases and the off-diagonal elements represent the cross-flexibilities. We also showed that in the case of a rigid-jointed structure each release load-pair can be associated with a "ring". Each leading diagonal element is formed by adding up the flexibilities of the members making up the corresponding ring, while the off-diagonal elements consist of the flexibilities of elements common to two rings. It follows that if we have two rings coupled by a very flexible member then the matrix $\mathbf{B}^t\mathbf{F}_m\mathbf{B}$ will be ill-conditioned in exactly the same way that the stiffness matrix $\mathbf{K}$ is ill-conditioned when two joints are coupled by a very stiff member.

We illustrate this by considering the simple example shown in Fig. 11.10.

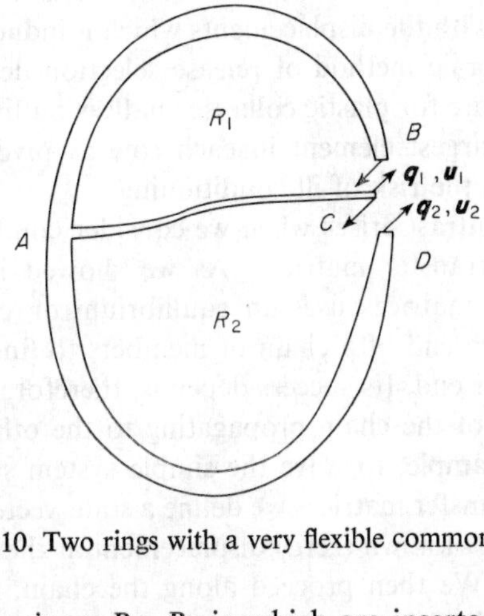

FIG. 11.10. Two rings with a very flexible common member.

Here we have two rings, R_1, R_2 in which are inserted releases. The gaps produced at the releases by the external loads are u_1, u_2 and we wish to find the load pairs q_1, q_2 which will reduce these gaps to zero. The equations for q_1 and q_2 will be of the general form

$$\begin{bmatrix} u_1 \\ u_2 \end{bmatrix} = \begin{bmatrix} F_{AB}' + F_{AC}' & -F_{AC}' \\ -F_{AC}' & F_{AD}' + F_{AC}' \end{bmatrix} \begin{bmatrix} q_1 \\ q_2 \end{bmatrix} \tag{11.11}$$

If now the member AC is very flexible compared with members AB and AD then F_{AC}' will be very much greater than F_{AB}' or F_{AD}' and equation (11.11) will be ill-conditioned in exactly the same way as equations (11.9) or (11.10).

Since the choice of releases in the compatibility method is normally under the control of the analyst it is desirable to have a general criterion as to what constitutes a "good" system of releases. As far as the conditioning of the flexibility matrix is concerned the main aim should be to make the leading diagonal elements as large as possible in comparison with the off-diagonal elements. Physically this corresponds to choosing a system of releases in which the direct flexibilities of the rings are large compared with the flexibilities of the members which couple the rings together. An alternative statement of this criterion is that the displacement which a release load-pair induces at its own release point should be as large as possible compared with the displacements which it induces at other release points. In the algebraic method of release selection described in Section 8.5 (and the procedure for plastic collapse analysis outlined in Section 7.7) the choice of the largest element in each row as pivotal element automatically minimizes the risk of ill-conditioning.

An interesting contrast arises when we consider conditioning problems in the method of transfer matrices. As we showed in Chapter 9, the method of transfer matrices uses an equilibrium or compatibility condition applied at one end of a chain of members to find unknowns associated with the other end. Its success depends, therefore, on a disturbance applied at one end of the chain propagating to the other without undue attenuation. For example, to solve the simple system shown in Fig. 11.7 by the method of transfer matrices we define a state-vector consisting of an unknown force and a known (zero) displacement at the extreme left-hand end of the system. We then proceed along the chain, finally applying a condition of zero displacement to the displacement component of the state-vector at the extreme right-hand end. If the stiffness K of the centre spring is very small compared with the stiffness k of the other springs this condition of zero displacement will be very little affected by anything which happens at the left-hand end of the system. In the extreme case where K is zero the transfer matrix method fails completely, since a disturbance can never propagate from one end of the system to the other. It follows, therefore, that the transfer matrix method encounters conditioning trouble in precisely the situation where the equilibrium method leads to well-conditioned equations. Conversely it may be satisfactory in situations where the application of the equilibrium method leads to ill-conditioned equations.

A Summary of the Main Equations

1. *Notation*

$\boldsymbol{r}$	is a vector of *independent* components, in terms of which the stress state of an element may be expressed.
$\boldsymbol{e}$	is the corresponding vector which defines the deformations. $\boldsymbol{r}$ and $\boldsymbol{e}$ satisfy the equation $\boldsymbol{r}^t\boldsymbol{e}^* =$ virtual work done on the element in a virtual deformation $\boldsymbol{e}^*$. For linear elastic elements $\boldsymbol{r} = \boldsymbol{K}\boldsymbol{e}$, $\boldsymbol{e} = \boldsymbol{F}\boldsymbol{r}$, where $\boldsymbol{F} = \boldsymbol{K}^{-1}$.
$\mathbf{r}$, $\mathbf{e}$	are the vectors formed by combining the vectors $\boldsymbol{r}$, $\boldsymbol{e}$ for all the elements. They are related by the equations $\mathbf{r} = \mathbf{K_m e}$, $\mathbf{e} = \mathbf{F_m r}$.
$\boldsymbol{p}_i$	are the nodal loads acting on an element, where $i = 1,2, \ldots,$ denotes the individual nodes.
$\boldsymbol{d}_i$	are the corresponding nodal displacements. $\boldsymbol{p}_i$ and $\boldsymbol{d}_i$ satisfy the equation $\boldsymbol{p}_i{}^t\boldsymbol{d}_i{}^* =$ virtual work done on the element in a virtual deformation $\boldsymbol{d}_i{}^*$. The set of nodal loads $\boldsymbol{p}_i$ must satisfy the element equilibrium equations $\boldsymbol{p}_i = \boldsymbol{H}_i\boldsymbol{r}$. For sufficiently small displacements the matrices $\boldsymbol{H}_i$ may be based on the geometry of the *undeformed* structure. The corresponding nodal displacements $\boldsymbol{d}_i$ are then related to the deformation vector $\boldsymbol{e}$ by the element compatibility equations $\boldsymbol{e} = \boldsymbol{H}_i{}^t\boldsymbol{d}_i$. A set $\boldsymbol{d}_i$ such that $\boldsymbol{H}_i{}^t\boldsymbol{d}_i = \boldsymbol{0}$ represents a (small) rigid-body displacement of the element.
$\boldsymbol{p}_i{}'$, $\boldsymbol{d}_i{}'$	are the corresponding vectors in global coordinates, where $\boldsymbol{p}_i{}' = \boldsymbol{T}\boldsymbol{p}_i$, $\boldsymbol{d}_i = \boldsymbol{T}^t\boldsymbol{d}_i{}'$.
$(\boldsymbol{p}_{\text{equiv}})_i$	(or $(\boldsymbol{p}'_{\text{equiv}})_i$ in global coordinates) are the nodal loads which are statically equivalent to the external loads acting on an element.

$\boldsymbol{p}_J$ denotes the external load acting at node J of a structure. It is equal to the concentrated external load at J plus appropriate loads $\boldsymbol{p}'_{\text{equiv}}$ contributed by elements which have J as one of their nodes. It is always expressed in global coordinates.

$\boldsymbol{d}_J$ denotes the displacement of node J in global coordinates. $\boldsymbol{p}_J$ and $\boldsymbol{d}_J$ satisfy the equation $\boldsymbol{p}_J{}^t\boldsymbol{d}_J{}^* =$ virtual work done by the external loads on the structure in a virtual displacement $\boldsymbol{d}_J{}^*$ (all other nodes remaining stationary).

p, d are the vectors formed by combining the vectors $\boldsymbol{p}_J$, $\boldsymbol{d}_J$ for all the nodes of a structure.

2. *The single finite element*

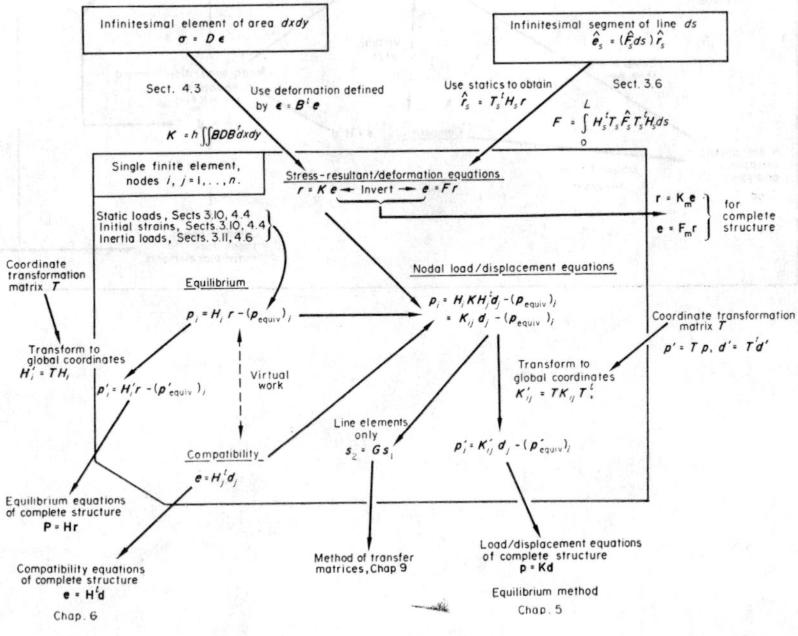

3. *The Equilibrium or Displacement method*

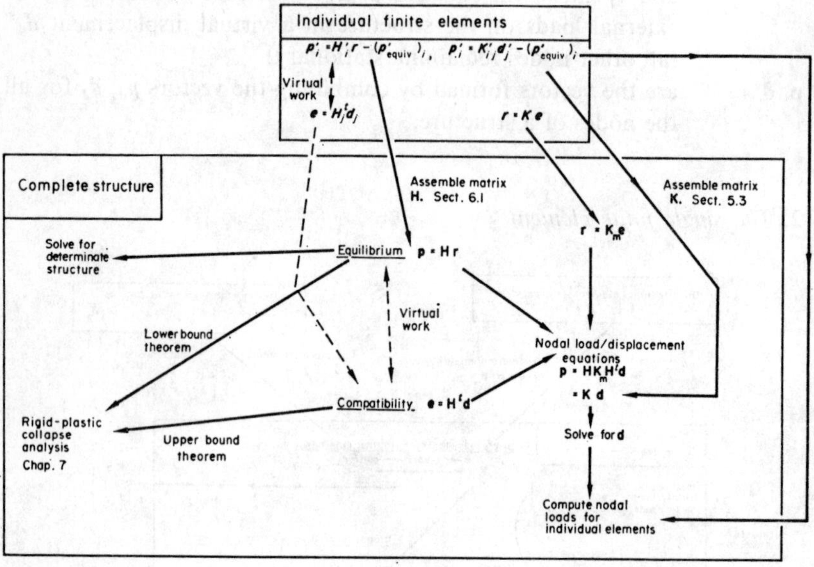

4. *The Compatibility or Force method*

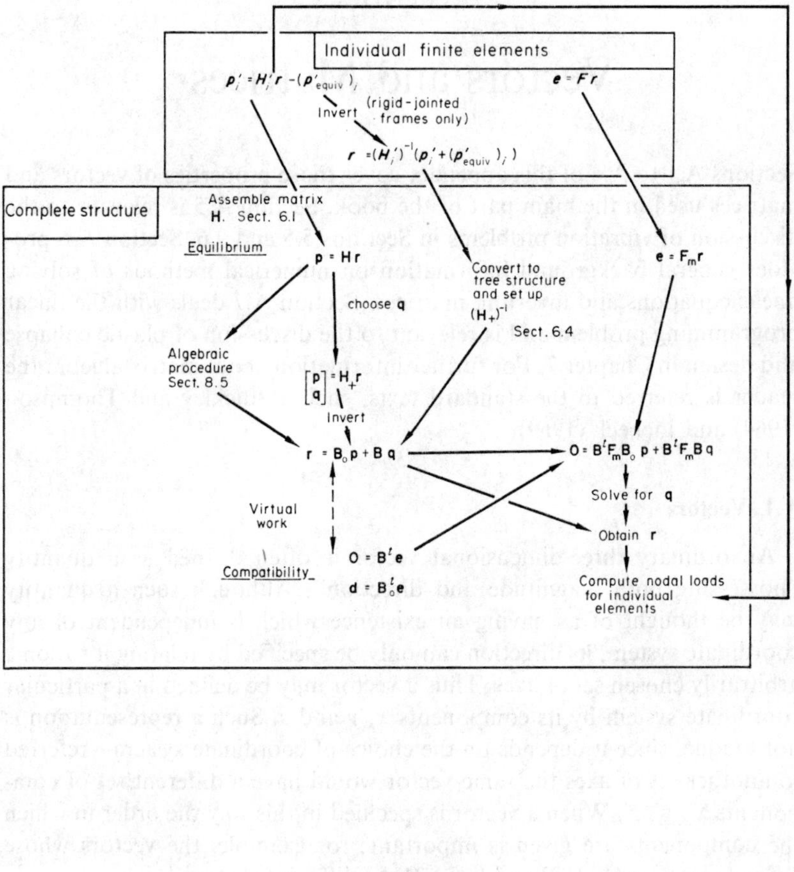

APPENDIX

Vectors and Matrices

Sections A.1 to A.4 of this appendix cover those properties of vectors and matrices used in the main part of the book. Section A.5 is relevant to the discussion of vibration problems in Sections 5.5 and 9.6. Section A.6 provides general background information on numerical methods of solving linear equations and inverting matrices. Section A.7 deals with the linear programming problem and is relevant to the discussion of plastic collapse and design in Chapter 7. For further information about matrix algebra the reader is referred to the standard texts, such as Bickley and Thompson (1964) and Liebeck (1969).

A.1. Vectors

An ordinary three-dimensional vector is often defined as a quantity "possessing both magnitude and direction". Although such a quantity may be thought of as having an existence which is independent of any coordinate system, its direction can only be specified by relating it to some arbitrarily chosen set of axes. Thus a vector may be defined in a particular coordinate system by its components x, y and z. Such a representation is not unique, since it depends on the choice of coordinate system—referred to another set of axes the same vector would have a different set of components x', y', z'. When a vector is specified in this way the order in which the components are given is important: for example, the vectors whose components are [1, 2, 3] and [3, 1, 2] are different vectors.

Conversely, any set of three numbers or variables arranged in a particular order may be regarded as defining a vector in the three-dimensional space formed by an appropriate set of coordinate axes. More generally, a set of n scalar quantities may be regarded as defining a vector in an n-dimensional space. This does not mean that the reader must attempt to visualize an n-dimensional space, but merely that many of the results which can be proved by geometrical argument in the case of three-dimensional vectors can be extended formally to n dimensions, and can be discussed in the same sort of geometrical language.

In this book a vector is defined as a set of scalar quantities $x_1, x_2, \ldots, x_n$, and is written for the sake of conciseness as x_i (where i is assumed to take all values from 1 to n) or $\boldsymbol{x}$. The individual scalar x's are referred to as the *components* of the vector. Where a physical quantity such as a load or a displacement is written as a vector the values of the components depend on the coordinate system in which these components are measured, so that essentially the same physical quantity will be represented by different vectors when viewed from different coordinate systems.

Vectors are normally written as columns of numbers or symbols, enclosed in square brackets. The rules for the manipulation of general n-dimensional vectors are very similar to those used in dealing with ordinary three-dimensional ones. Thus two vectors are equal only if all their components are equal, while the *sum* of two vectors is formed by adding corresponding components. By analogy with the normal three-dimensional definition, the *length* of a general vector $\boldsymbol{x}$ is defined to be

$$\sqrt{\left(\sum_{i=1}^{n} x_i^2\right)}.$$

Multiplication of a vector by a scalar merely increases each component in the same proportion, or, in geometrical terms, changes the "length" of the vector without altering its "direction". Thus

$$k\boldsymbol{x} = k\begin{bmatrix} x_1 \\ x_2 \\ \cdot \\ \cdot \\ x_n \end{bmatrix} = \begin{bmatrix} kx_1 \\ kx_2 \\ \cdot \\ \cdot \\ kx_n \end{bmatrix}$$

If two vectors $a_1, a_2, \ldots, a_n$ and $b_1, b_2, \ldots, b_n$ have the same number of components their *scalar product* is defined to be the scalar quantity

$c = a_1b_1 + a_2b_2 + \ldots a_nb_n$, which may be written more concisely as

$$c = \sum_{i=1}^{n} a_ib_i.$$

It is common practice to drop the summation symbol in such an expression and write it simply as $c = a_ib_i$, the repetition of the suffix i being regarded as implying summation over all its possible values. It should be noted that summation is only implied when the repetition of the suffix occurs within a product—the equation $a_i = x_i$, for example, does not imply any summation. In ordinary three-dimensional vector algebra it can easily be shown that two vectors whose scalar product is zero must be at right angles to each other. In the general case two vectors whose scalar product is zero are said to be *orthogonal*.

The equation

$$a_1x_1 + a_2x_2 + a_3x_3 = 0 \tag{A.1}$$

represents a plane through the origin (i.e. a two-dimensional space) lying in the three-dimensional space defined by the axes x_1, x_2, x_3, and all vectors whose components x_1, x_2, x_3 have to satisfy (A.1) must lie in this plane. Although such vectors have three components only two can be specified arbitrarily, since the third is then determined by (A.1). In such a case the three components are said to be *linearly dependent*. This idea may be extended to n-dimensional vectors. If the components of an n-dimensional vector $\boldsymbol{x}$ have to satisfy an equation of the form

$$a_1x_1 + a_2x_2 + \ldots a_nx_n = 0 \tag{A.2}$$

then the components are said to be linearly dependent, since only $n - 1$ of them can be specified arbitrarily. Equation (A.2) restricts the vector $\boldsymbol{x}$ to a space of $n - 1$ dimensions, sometimes called a *hyperplane*, lying in the n-dimensional space formed by the axes $x_1, x_2, \ldots, x_n$.

A.2. Matrices

In the analysis of linear systems there are many occasions when one vector is expressed as a linear function of another. That is, a set of variables $x_1, \ldots, x_n$ is related to another set of variables $y_1, \ldots, y_m$ by a system of linear equations.

$$\left.\begin{aligned} x_1 &= a_{11}y_1 + a_{12}y_2 + \ldots a_{1m}y_m \\ x_2 &= a_{21}y_1 + a_{22}y_2 + \ldots a_{2m}y_m \\ &\cdot \quad \cdot \quad \cdot \quad \cdot \quad \cdot \\ &\cdot \quad \cdot \quad \cdot \quad \cdot \quad \cdot \\ x_n &= a_{n1}y_1 + a_{n2}y_2 + \ldots a_{nm}y_m \end{aligned}\right\} \tag{A.3}$$

where the coefficients a_{11}, a_{12}, etc., are constant. In these equations the first suffix of each coefficient defines the row in which it stands, while the second defines the column. Equations (A.3) may be written in condensed form as

$$x_i = \sum_{j=1}^{m} a_{ij}y_j \quad (i = 1, \ldots, n)$$

or, using the summation convention mentioned in Section A.1, simply as

$$x_i = a_{ij}y_j. \tag{A.4}$$

Equation (A.4) can be given rather a different interpretation. The symbols x_i and y_j can be thought of as representing the *complete* vectors $\boldsymbol{x}$ and $\boldsymbol{y}$ rather than particular *individual* components, with the symbol a_{ij} representing the whole set of $n \times m$ coefficients appearing in (A.3). The set a_{ij} is called a *matrix* and is written in square brackets. Thus (A.3) becomes

$$\begin{bmatrix} x_1 \\ x_2 \\ \cdot \\ \cdot \\ x_n \end{bmatrix} = \begin{bmatrix} a_{11} & a_{12} & \cdot & \cdot & \cdot & a_{1m} \\ a_{21} & a_{22} & \cdot & \cdot & \cdot & a_{2m} \\ \cdot & \cdot & \cdot & \cdot & \cdot & \cdot \\ \cdot & \cdot & \cdot & \cdot & \cdot & \cdot \\ a_{n1} & a_{n2} & \cdot & \cdot & \cdot & a_{nm} \end{bmatrix} \begin{bmatrix} y_1 \\ y_2 \\ \cdot \\ y_m \end{bmatrix} \tag{A.5}$$

or, more briefly,

$$\boldsymbol{x} = A\boldsymbol{y}. \tag{A.6}$$

Equation (A.6) may be interpreted as stating that the vector $\boldsymbol{x}$ is equal to the vector $\boldsymbol{y}$ "multiplied" by the matrix A. Since these symbols do not now represent simple algebraic quantities, it is necessary to define the meaning of the word "multiplication" in this context. This is done by making equations (A.3) . . . (A.6) all identical in meaning. In other words, the multiplication of a vector by a matrix involves taking the scalar product of each row of the matrix with the vector. It is obvious that for such a multiplication to be possible the number of *columns* in the matrix must be equal to the number of *components* in the vector: the result of the multiplication is a vector with as many components as there are *rows* in the matrix. For example, the linear relation

$$x_1 = 3y_1 - 2y_2 + 4y_3,$$
$$x_2 = 2y_1 + 3y_2 - y_3$$

appears in matrix form as

$$\begin{bmatrix} x_1 \\ x_2 \end{bmatrix} = \begin{bmatrix} 3 & -2 & 4 \\ 2 & 3 & -1 \end{bmatrix} \begin{bmatrix} y_1 \\ y_2 \\ y_3 \end{bmatrix}$$

A matrix with an equal number of rows and columns is termed a *square* matrix, and the coefficients $a_{11}, a_{22}, \ldots a_{nn}$ are referred to collectively as the *leading diagonal.* A square matrix in which the only non-zero terms are on the leading diagonal is called a *diagonal* matrix, and if all these coefficients are equal to unity it is called a *unit* matrix, normally denoted by the symbol $\boldsymbol{I}$. A unit matrix clearly satisfies the equation $\boldsymbol{x} = \boldsymbol{I}\boldsymbol{x}$ for an arbitrary vector $\boldsymbol{x}$. A matrix entirely composed of zero coefficients is termed a *null* matrix, and is written as $\boldsymbol{0}$.

The next step is to consider what is meant by the "multiplication" of two matrices. Consider the two sets of equations

$$\begin{aligned} x_1 &= a_{11}y_1 + a_{12}y_2 + a_{13}y_3 \qquad & y_1 &= b_{11}z_1 + b_{12}z_2 \\ x_2 &= a_{21}y_1 + a_{22}y_2 + a_{23}y_3 \qquad & y_2 &= b_{21}z_1 + b_{22}z_2 \\ & & y_3 &= b_{31}z_1 + b_{32}z_2 \end{aligned} \tag{A.7}$$

In matrix notation these become

$$\begin{bmatrix} x_1 \\ x_2 \end{bmatrix} = \begin{bmatrix} a_{11} & a_{12} & a_{13} \\ a_{21} & a_{22} & a_{23} \end{bmatrix} \begin{bmatrix} y_1 \\ y_2 \\ y_3 \end{bmatrix}, \quad \begin{bmatrix} y_1 \\ y_2 \\ y_3 \end{bmatrix} = \begin{bmatrix} b_{11} & b_{12} \\ b_{21} & b_{22} \\ b_{31} & b_{32} \end{bmatrix} \begin{bmatrix} z_1 \\ z_2 \end{bmatrix}$$

and eliminating the vector $\boldsymbol{y}$ gives the formal result

$$\begin{bmatrix} x_1 \\ x_2 \end{bmatrix} = \begin{bmatrix} a_{11} & a_{12} & a_{13} \\ a_{21} & a_{22} & a_{23} \end{bmatrix} \begin{bmatrix} b_{11} & b_{12} \\ b_{21} & b_{22} \\ b_{31} & b_{32} \end{bmatrix} \begin{bmatrix} z_1 \\ z_2 \end{bmatrix} \quad \text{(A.8)}$$

Alternatively y_1, y_2, y_3 may be eliminated from equations (A.7) by the ordinary rules of algebra to give

$$x_1 = (a_{11}b_{11} + a_{12}b_{21} + a_{13}b_{31})z_1 + (a_{11}b_{12} + a_{12}b_{22} + a_{13}b_{32})z_2$$

$$x_2 = (a_{21}b_{11} + a_{22}b_{21} + a_{23}b_{31})z_1 + (a_{21}b_{12} + a_{22}b_{22} + a_{23}b_{32})z_2$$

which may be written in matrix form as

$$\begin{bmatrix} x_1 \\ x_2 \end{bmatrix} = \begin{bmatrix} a_{11}b_{11} + a_{12}b_{21} + a_{13}b_{31} & a_{11}b_{12} + a_{12}b_{22} + a_{13}b_{32} \\ a_{21}b_{11} + a_{22}b_{21} + a_{23}b_{31} & a_{21}b_{12} + a_{22}b_{22} + a_{23}b_{32} \end{bmatrix} \begin{bmatrix} z_1 \\ z_2 \end{bmatrix} \quad \text{(A.9)}$$

Comparing (A.8) and (A.9) gives immediately

$$\begin{bmatrix} a_{11} & a_{12} & a_{13} \\ a_{21} & a_{22} & a_{23} \end{bmatrix} \begin{bmatrix} b_{11} & b_{12} \\ b_{21} & b_{22} \\ b_{31} & b_{32} \end{bmatrix} = \begin{bmatrix} a_{11}b_{11} + a_{12}b_{21} + a_{13}b_{31} & a_{11}b_{12} + a_{12}b_{22} + a_{13}b_{32} \\ a_{21}b_{11} + a_{22}b_{21} + a_{23}b_{31} & a_{21}b_{12} + a_{22}b_{22} + a_{23}b_{32} \end{bmatrix}$$

Thus the product of two matrices $\boldsymbol{A}$ and $\boldsymbol{B}$ is formed by taking the scalar product of each *row* of $\boldsymbol{A}$ with each *column* of $\boldsymbol{B}$. More explicitly, if $\boldsymbol{C} = \boldsymbol{AB}$ then the element in the ith row and the jth column of $\boldsymbol{C}$ is obtained by taking the scalar product of the ith row of $\boldsymbol{A}$ and the jth column of $\boldsymbol{B}$. It follows that for such a multiplication to be possible the number of *columns* in $\boldsymbol{A}$ must equal the number of *rows* in $\boldsymbol{B}$. It is always the rows of the first matrix which multiply the columns of the second, so that the product $\boldsymbol{BA}$ is not necessarily equal to $\boldsymbol{AB}$, and indeed may not exist at all. When referring to a matrix product such as $\boldsymbol{AB}$ the order of multiplication may be specified by saying that $\boldsymbol{A}$ "pre-multiplies" $\boldsymbol{B}$, or that $\boldsymbol{B}$ "post-multiplies" $\boldsymbol{A}$.

Every matrix A has associated with it another matrix called its *transpose*, written as A^t. Each *row* of A is equal to the corresponding *column* of A^t. A matrix for which $A^t = A$ is called a *symmetric* matrix. Such a matrix is necessarily square and the elements satisfy the condition $a_{ij} = a_{ji}$. It is easy to verify that the product of two symmetric matrices is not necessarily symmetric.

A vector may be regarded as a matrix containing only one row or column. The scalar product of two vectors $\boldsymbol{a}$ and $\boldsymbol{b}$ may therefore be written in the form

$$[a_1 \quad a_2 \quad . \quad . \quad . \, a_n]\begin{bmatrix} b_1 \\ b_2 \\ . \\ . \\ . \\ b_n \end{bmatrix} = c \tag{A.10}$$

where the vector $\boldsymbol{a}$ is written as a row rather than a column in order to agree with the general rule for matrix multiplication. It is important to realize that such a "row vector" is not essentially different from the same symbols written as a column. It is simply a matter of arranging the components on paper in such a way that they conform to the rule for multiplication. Since a row vector is the transpose of the corresponding column vector, (A.10) may be written as $\boldsymbol{a}^t\boldsymbol{b} = c$, and it is obvious that $\boldsymbol{a}^t\boldsymbol{b} = \boldsymbol{b}^t\boldsymbol{a}$. More generally, it is easy to show that for matrices $(\boldsymbol{AB})^t = \boldsymbol{B}^t\boldsymbol{A}^t$ and that $(\boldsymbol{AB} \ldots \boldsymbol{N})^t = \boldsymbol{N}^t \ldots \boldsymbol{B}^t\boldsymbol{A}^t$.

Other properties of matrices can be derived by considering operations carried out on linear systems of equations. For example, if

$$\begin{aligned} x_1 &= 4y_1 + 2y_2 \qquad & z_1 &= 2y_1 + 6y_2, \\ x_2 &= 3y_1 + 5y_2 \qquad & z_2 &= 4y_1 - y_2, \end{aligned} \tag{A.11}$$

then it follows that

$$\begin{aligned} x_1 + z_1 &= (4 + 2)y_1 + (2 + 6)y_2, \\ x_2 + z_2 &= (3 + 4)y_1 + (5 - 1)y_2. \end{aligned} \tag{A.12}$$

In matrix notation (A.11) may be written as $\boldsymbol{x} = \boldsymbol{Ay}$, $\boldsymbol{z} = \boldsymbol{By}$, and (A.12) as $\boldsymbol{x} + \boldsymbol{z} = (\boldsymbol{A} + \boldsymbol{B})\boldsymbol{y} = \boldsymbol{Cy}$, where $\boldsymbol{C}$, the *sum* of $\boldsymbol{A}$ and $\boldsymbol{B}$, is formed by adding corresponding elements of the two matrices. It follows that one cannot add two matrices unless they have the same number of rows and the same number of columns. In the same way it is possible to show that the associative laws of algebra

$$\boldsymbol{A}(\boldsymbol{B} + \boldsymbol{C}) = \boldsymbol{AB} + \boldsymbol{AC}, \quad (\boldsymbol{AB})\boldsymbol{C} = \boldsymbol{A}(\boldsymbol{BC})$$

hold for matrices, although the commutative law does not.

In structural analysis it is often convenient to use matrices in which the elements are themselves smaller matrices. Thus the equations $\boldsymbol{Ax} + \boldsymbol{By} = \boldsymbol{a}$, $\boldsymbol{Cx} + \boldsymbol{Dy} = \boldsymbol{b}$ may be written in the form

$$\begin{bmatrix} \boldsymbol{A} & \boldsymbol{B} \\ \boldsymbol{C} & \boldsymbol{D} \end{bmatrix} \begin{bmatrix} \boldsymbol{x} \\ \boldsymbol{y} \end{bmatrix} = \begin{bmatrix} \boldsymbol{a} \\ \boldsymbol{b} \end{bmatrix}$$

A matrix which is divided into blocks of coefficients in this way is said to be *partitioned*, and the elements $\boldsymbol{A}$, $\boldsymbol{B}$, etc., are called *submatrices.*

An important quantity associated with a symmetric matrix is its *quadratic form.* The quadratic form associated with the matrix $\boldsymbol{A}$ is simply the scalar function $\boldsymbol{x}^t\boldsymbol{Ax}$, where $\boldsymbol{x}$ is an arbitrary vector. Written out in suffix notation this takes the form $a_{ij}x_ix_j$, summation being implied over both i and j. A matrix whose quadratic form is always strictly positive for an arbitrary non-zero vector $\boldsymbol{x}$ is called *positive-definite.* The quadratic form associated with a symmetric matrix often has a direct physical significance. For example, if the displacements $\boldsymbol{d}$ of a structure are related to the external loads $\boldsymbol{p}$ by the linear equation $\boldsymbol{p} = \boldsymbol{Kd}$, then the work done on the structure during the application of the loads is $\boldsymbol{d}^t\boldsymbol{p}/2 = \boldsymbol{d}^t\boldsymbol{Kd}/2$. Thus the quadratic form associated with the matrix $\boldsymbol{K}$ is proportional to the strain energy of the structure. If $\boldsymbol{K}$ is positive-definite then the strain energy is always positive and the structure is stable.

A.3. Matrix inversion

The equation

$$\boldsymbol{y} = \boldsymbol{Ax} \tag{A.13}$$

may be thought of as having the effect of transforming an arbitrary vector $\boldsymbol{x}$ into another vector $\boldsymbol{y}$ and is often described as a linear transformation. In two dimensions the vector $\boldsymbol{x}$ may be regarded as a line drawn on a rubber sheet, the multiplication of $\boldsymbol{x}$ by $\boldsymbol{A}$ corresponding to a stretching of the sheet in a uniform manner. The following are examples of simple two-dimensional transformations:

1. $\boldsymbol{A} = \begin{bmatrix} 1 & 0 \\ 0 & 1 \end{bmatrix}$ $\boldsymbol{A}$ is a unit matrix: this transformation leaves every vector unchanged.

2. $\boldsymbol{A} = \begin{bmatrix} a & 0 \\ 0 & b \end{bmatrix}$ $\boldsymbol{A}$ is a diagonal matrix: this transformation multiplies the x_1 component of any vector by a factor a, and the x_2 component by a factor b.

3. $\boldsymbol{A} = \begin{bmatrix} \cos\alpha & -\sin\alpha \\ \sin\alpha & \cos\alpha \end{bmatrix}$ This produces a rotation of any vector through an angle α, as shown in Fig. A.1.

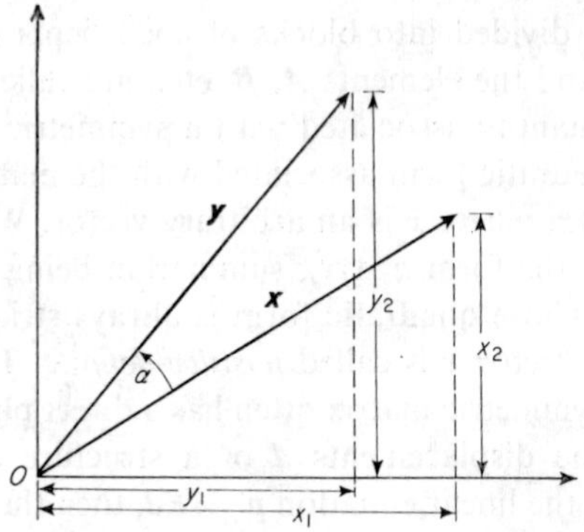

FIG. A.1. Rotation of a two-dimensional vector through an angle α.

In these three examples any particular vector $\boldsymbol{x}$ gives a unique vector $\boldsymbol{y}$, and conversely a choice of $\boldsymbol{y}$ defines a corresponding unique vector $\boldsymbol{x}$. This implies that an inverse transformation exists of the form

$$\boldsymbol{x} = \boldsymbol{B}\boldsymbol{y} \tag{A.14}$$

where $\boldsymbol{B}$ is some function of $\boldsymbol{A}$. The matrix $\boldsymbol{B}$ is called the *inverse* of $\boldsymbol{A}$ and is written $\boldsymbol{A}^{-1}$. Methods of computing the inverse of a matrix will be found in any standard textbook on matrix algebra—an elementary method is described in Section A.6.

The elimination of $\boldsymbol{y}$ from equations (A.13) and (A.14) gives $\boldsymbol{x} = \boldsymbol{A}^{-1}\boldsymbol{A}\boldsymbol{x}$. Since the only matrix which leaves all vectors $\boldsymbol{x}$ unchanged is the unit matrix $\boldsymbol{I}$, it follows that $\boldsymbol{A}^{-1}\boldsymbol{A} = \boldsymbol{I}$, just as $a^{-1}a = 1$ for ordinary scalars. This result is useful in checking the computed value of an inverse matrix.

For a matrix to possess an inverse it must necessarily be square, but this alone is not sufficient. Consider, for example, the transformation

$$\begin{bmatrix} y_1 \\ y_2 \end{bmatrix} = \begin{bmatrix} \cot \alpha & 1 \\ 1 & \tan \alpha \end{bmatrix} \begin{bmatrix} x_1 \\ x_2 \end{bmatrix} \tag{A.15}$$

where α is constant. It is apparent that, irrespective of the values of x_1 and x_2, the components of the vector $\boldsymbol{y}$ are not independent, but must satisfy the equation

$$y_2 = y_1 \tan \alpha. \tag{A.16}$$

Thus (A.15) transforms any vector $\boldsymbol{x}$ into a vector lying in a certain direction. In fact all vectors $\boldsymbol{x}$ for which the function $x_1 + x_2 \tan \alpha$ is constant transform into the same vector $\boldsymbol{y}$ as shown in Fig. A.2. In such a case it is clearly impossible, given a particular vector $\boldsymbol{y}$, to find a unique corresponding vector $\boldsymbol{x}$, so that the inverse transformation (A.14) does not exist. A transformation or a matrix which has no inverse is said to be *singular*.

In general a singular transformation is one in which the resulting vector $\boldsymbol{y}$ has a smaller number of independent components than the original vector $\boldsymbol{x}$. In geometrical language this is equivalent to saying that the vector $\boldsymbol{y}$ lies in a space with a smaller number of dimensions than that of the space associated with the vector $\boldsymbol{x}$. This may be because the matrix $\boldsymbol{A}$ is a rectangular matrix with fewer rows than columns, so that the vector $\boldsymbol{y}$ necessarily has fewer components than $\boldsymbol{x}$. It may, on the other hand, be due to the fact that although $\boldsymbol{x}$ and $\boldsymbol{y}$ have the same number of components, the square matrix $\boldsymbol{A}$ is such that the components of $\boldsymbol{y}$ have to satisfy one or more conditions such as (A.16). As mentioned in Section A.1, if the components of $\boldsymbol{y}$ satisfy (independently of the choice of $\boldsymbol{x}$) one or more conditions of the form

$$c_1y_1 + c_2y_2 + \ldots c_ny_n = 0 \tag{A.17}$$

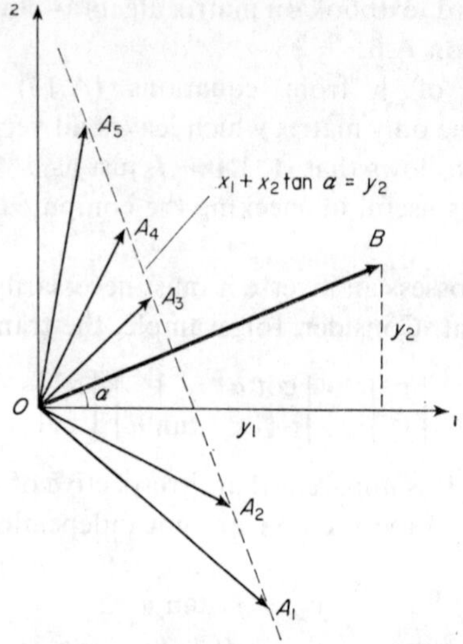

FIG. A.2. A singular transformation: all the vectors OA_1 OA_2 . . ., transform into the same vector OB.

then it is impossible to give arbitrary values to all the components, so that the vector $\boldsymbol{y}$ does not, in fact, occupy a space of the same number of dimensions as $\boldsymbol{x}$. In these circumstances the inverse transformation (A.14) does not exist. Equation (A.13) may be written in the form

$$
\begin{aligned}
y_1 &= \boldsymbol{a}_1{}^t\boldsymbol{x} \\
y_2 &= \boldsymbol{a}_2{}^t\boldsymbol{x} \\
&\vdots \\
y_n &= \boldsymbol{a}_n{}^t\boldsymbol{x}
\end{aligned}
$$

where $\boldsymbol{a}_1{}^t, \ldots, \boldsymbol{a}_n{}^t$ are the successive rows of the matrix $\boldsymbol{A}$. Substituting these expressions in (A.17) gives $(c_1\boldsymbol{a}_1{}^t + c_2\boldsymbol{a}_2{}^t + \ldots c_n\boldsymbol{a}_n{}^t)\boldsymbol{x} = 0$, and since $\boldsymbol{x}$ may be any vector it follows that

$$c_1\boldsymbol{a}_1 + c_2\boldsymbol{a}_2 + \ldots c_n\boldsymbol{a}_n = \boldsymbol{0}. \tag{A.18}$$

Thus a square matrix is singular if its rows satisfy one or more equations such as (A.18), i.e. if one or more of the rows are merely linear combinations of the others. If no such relationship exists the rows of the matrix are said to be *linearly independent* and the matrix is then non-singular. The reader familiar with the theory of determinants will recognize that a matrix is singular if and only if its determinant is zero.

It is instructive to arrive at the same result by a slightly different line of reasoning. Consider the set of three simultaneous equations

$$a_{11}x_1 + a_{12}x_2 + a_{13}x_3 = b_1,$$
$$a_{21}x_1 + a_{22}x_2 + a_{23}x_3 = b_2,$$
$$a_{31}x_1 + a_{32}x_2 + a_{33}x_3 = b_3,$$

which in matrix form is equivalent to

$$\boldsymbol{Ax} = \boldsymbol{b} \tag{A.19}$$

Equation (A.19) may be written

$$\left.\begin{aligned} \boldsymbol{a}_1{}^t\boldsymbol{x} &= b_1,\\ \boldsymbol{a}_2{}^t\boldsymbol{x} &= b_2,\\ \boldsymbol{a}_3{}^t\boldsymbol{x} &= b_3, \end{aligned}\right\} \tag{A.20}$$

where $\boldsymbol{a}_1{}^t$, $\boldsymbol{a}_2{}^t$, $\boldsymbol{a}_3{}^t$ are the rows of the matrix $\boldsymbol{A}$. Each of these vectors may be made of unit length by dividing the original equations by the appropriate factors. If this is done, then (A.20) states that b_1, b_2, b_3 are the components of the vector $\boldsymbol{x}$ in the directions of the unit vectors $\boldsymbol{a}_1$, $\boldsymbol{a}_2$, $\boldsymbol{a}_3$. This information will be sufficient to determine $\boldsymbol{x}$, provided that these three unit vectors constitute a set of axes in which it is possible to define a general three-dimensional vector. This will be so provided that the vectors $\boldsymbol{a}_1$, $\boldsymbol{a}_2$, $\boldsymbol{a}_3$ are not coplanar, which is equivalent to saying that they do not satisfy an equation of the form

$$c_1\boldsymbol{a}_1 + c_2\boldsymbol{a}_2 + c_3\boldsymbol{a}_3 = \boldsymbol{0}.$$

This is the same condition as before.

This analysis leads to the idea of *ill-conditioned* systems of equations. If the vectors $\boldsymbol{a}_1$, $\boldsymbol{a}_2$, $\boldsymbol{a}_3$ are not coplanar, then theoretically the inverse matrix $\boldsymbol{A}^{-1}$ exists, and it is possible to solve (A.19) for the vector $\boldsymbol{x}$.

However, if the vector a_3 lies very nearly in the plane defined by a_1 and a_2 then the three unit vectors obviously form a "bad" set of axes for describing a three-dimensional space, and considerable numerical difficulty may be experienced in obtaining an accurate solution of the equations. In such a case the matrix A is said to be ill-conditioned.

In contrast, the "best" sets of axes for describing a three-dimensional space are rectangular cartesian ones, in which the base vectors a_1, a_2, a_3 are mutually perpendicular. This condition may be written in the form $a_i^t a_j = 0$, $(i \neq j)$. A matrix in which the rows form a system of mutually orthogonal unit vectors is termed an *orthogonal* matrix. Since for an orthogonal matrix

$$a_i^t a_j = 0 \quad (i \neq j),$$

$$a_i^t a_j = 1 \quad (i = j),$$

it follows that $A^t A = I$, so that the inverse of an orthogonal matrix is equal to its transpose.

It is easy to show that if each of the matrices A, B, and C is non-singular then $(ABC)^{-1} = C^{-1}B^{-1}A^{-1}$. This result may be proved by pre-multiplying both sides of the equation by ABC, giving $(ABC)(ABC)^{-1} = ABCC^{-1}B^{-1}A^{-1}$. Both sides of the latter equation are clearly equal to the unit matrix.

A.4. Coordinate transformations

As mentioned in Section A.1, the definition of a vector as a set of components implies the existence of a coordinate system in which these components are measured. If in the equation

$$y = Ax \tag{A.21}$$

the symbols x and y stand for physical quantities such as loads and displacements, then this equation represents a linear relationship between these quantities as seen from the standpoint of one particular frame of reference. From a different reference frame the same relationship appears in a different form as $y' = A'x'$. Thus the matrix A is not an invariant mathematical description of a relationship between two physical quantities, but depends on the coordinate system used to define their components.

The following analysis shows how $\boldsymbol{A}$ and $\boldsymbol{A}'$ are related. For the sake of simplicity $\boldsymbol{x}$ and $\boldsymbol{y}$ are assumed to be ordinary three-dimensional vectors, whose components are initially defined in a given set of rectangular cartesian coordinates. Let $\boldsymbol{u}_1, \boldsymbol{u}_2, \boldsymbol{u}_3$ be a set of three mutually perpendicular unit vectors whose components (i.e. direction cosines) are defined relative to this set of axes. These three unit vectors may be thought of as defining a new system of coordinates, and in this new system the vector $\boldsymbol{x}$ has components $\boldsymbol{u}_1{}^t\boldsymbol{x}$, $\boldsymbol{u}_2{}^t\boldsymbol{x}$, $\boldsymbol{u}_3{}^t\boldsymbol{x}$. These components may be written as

$$\left.\begin{aligned} x_1' &= \boldsymbol{u}_1{}^t\boldsymbol{x}, \\ x_2' &= \boldsymbol{u}_2{}^t\boldsymbol{x}, \\ x_3' &= \boldsymbol{u}_3{}^t\boldsymbol{x}, \end{aligned}\right\} \tag{A.22}$$

or more concisely as

$$\boldsymbol{x}' = \boldsymbol{U}\boldsymbol{x} \tag{A.23}$$

where $\boldsymbol{U}$ is the orthogonal matrix whose rows are $\boldsymbol{u}_1{}^t$, $\boldsymbol{u}_2{}^t$, $\boldsymbol{u}_3{}^t$. Since $\boldsymbol{U}$ is orthogonal its inverse is equal to its transpose, so that

$$\boldsymbol{x} = \boldsymbol{U}^t\boldsymbol{x}'. \tag{A.24}$$

Equation (A.23) may also be used to find the vector $\boldsymbol{y}'$ corresponding to the vector $\boldsymbol{y}$, i.e. $\boldsymbol{y}' = \boldsymbol{U}\boldsymbol{y}$. Combining this equation with (A.21) and (A.24) gives

$$\boldsymbol{y}' = \boldsymbol{U}\boldsymbol{y} = \boldsymbol{U}\boldsymbol{A}\boldsymbol{x} = (\boldsymbol{U}\boldsymbol{A}\boldsymbol{U}^t)\boldsymbol{x}'.$$

Thus the relationship which was defined by the matrix $\boldsymbol{A}$ in the old coordinate system is defined by the matrix $\boldsymbol{A}' = \boldsymbol{U}\boldsymbol{A}\boldsymbol{U}^t$ in the new one. The matrices $\boldsymbol{A}$ and $\boldsymbol{A}'$ are said to be *congruent*. It is easy to show that if $\boldsymbol{A}$ is symmetric then so is $\boldsymbol{A}'$.

A.5. The principal axes problem

Transformations of the type described in the previous section may be carried out on any square matrix, whether symmetric or not. However, many of the matrices which occur in structural analysis are symmetric and, as such, possess a number of special properties.

The previous section showed how a relationship $\boldsymbol{y} = \boldsymbol{A}\boldsymbol{x}$ appears in another coordinate system as $\boldsymbol{y}' = \boldsymbol{A}'\boldsymbol{x}'$, where $\boldsymbol{A}'$ is derived from $\boldsymbol{A}$ by the congruent transformation $\boldsymbol{A}' = \boldsymbol{U}\boldsymbol{A}\boldsymbol{U}^t$. If $\boldsymbol{A}$ is symmetric it is possible to find an orthogonal transformation matrix $\boldsymbol{U}$ which makes $\boldsymbol{A}'$ purely diagonal, so that the relationship becomes

$$\begin{bmatrix} y_1' \\ y_2' \\ \cdot \\ \cdot \\ \cdot \\ \cdot \\ y_n' \end{bmatrix} = \begin{bmatrix} \lambda_1 & 0 & \cdot & \cdot & \cdot & \cdot & 0 \\ 0 & \lambda_2 & \cdot & \cdot & \cdot & \cdot & 0 \\ \cdot & \cdot & \cdot & \cdot & \cdot & \cdot & \cdot \\ \cdot & \cdot & \cdot & \cdot & \cdot & \cdot & \cdot \\ \cdot & \cdot & \cdot & \cdot & \cdot & \cdot & \cdot \\ \cdot & \cdot & \cdot & \cdot & \cdot & \cdot & \cdot \\ 0 & \cdot & \cdot & \cdot & \cdot & 0 & \lambda_n \end{bmatrix} \begin{bmatrix} x_1' \\ x_2' \\ \cdot \\ \cdot \\ \cdot \\ \cdot \\ x_n' \end{bmatrix}$$

all the λ's being real. (The transformation is also possible if $\boldsymbol{A}$ is not symmetric, but the λ's may be complex.) This equation may be written as

$$\begin{aligned} y_1' &= \lambda_1 x_1' \\ y_2' &= \lambda_2 x_2' \\ &\cdot \\ &\cdot \\ y_n' &= \lambda_n x_n' \end{aligned}$$

In other words, any linear relationship defined by a symmetric matrix is essentially one in which (in the appropriate coordinate system) individual components are simply multiplied by the appropriate λ's, so that y_1' depends only on x_1', y_2' only on x_2', etc. The unit vectors $\boldsymbol{u}_1, \boldsymbol{u}_2, \ldots, \boldsymbol{u}_n$ which form the axes of the new system of coordinates are called the *principal axes*, *characteristic vectors*, *eigenvectors*, or *latent vectors* of the matrix $\boldsymbol{A}$. The λ's which form the diagonal elements of $\boldsymbol{A}'$ are called the *eigenvalues* or *latent roots* of $\boldsymbol{A}$.

The eigenvalues and eigenvectors of a matrix may be found as follows. Consider the application of the linear transformation (A.21) to a vector $\boldsymbol{x}$ which is in the direction of one of the eigenvectors. This vector has only one non-zero component in the coordinate system defined by the $\boldsymbol{u}_i$'s, so that the effect of the transformation is to produce a vector $\boldsymbol{y}$ in the same direction as $\boldsymbol{x}$. Thus $\boldsymbol{x}$ satisfies the equation $\boldsymbol{A}\boldsymbol{x} = \lambda\boldsymbol{x}$, where λ is the appropriate eigenvalue. This equation may be written as

$$(\boldsymbol{A} - \lambda \boldsymbol{I})\boldsymbol{x} = \boldsymbol{0}. \tag{A.25}$$

Equation (A.25) implies that the rows of the matrix $(\boldsymbol{A} - \lambda \boldsymbol{I})$ are not linearly independent. It follows that the matrix is singular and consequently that the determinant $|\boldsymbol{A} - \lambda \boldsymbol{I}|$ is zero. Expansion of this determinant gives a polynomial of degree n in λ, and it can be shown that if $\boldsymbol{A}$ is symmetric then all the n roots of this polynomial are real. These n roots are the n eigenvalues $\lambda_1, \lambda_2, \ldots, \lambda_n$. In the simple case where the n roots are distinct, the eigenvector associated with any particular eigenvalue λ_i may be found by substituting this value into (A.25). This gives a set of n simultaneous equations of the form

$$\boldsymbol{B}_i \boldsymbol{x} = \boldsymbol{0} \tag{A.26}$$

where $\boldsymbol{B}_i = \boldsymbol{A} - \lambda_i \boldsymbol{I}$. Although $\boldsymbol{B}$ is singular it can be shown that if λ_i is a simple root then $n - 1$ of the simultaneous equations (A.26) are linearly independent. It is possible, therefore, to specify one component of the solution arbitrarily and then solve the equations for the remainder. This arbitrary specification of one component is equivalent to defining the length of the eigenvector, which is normally chosen to be unity. Thus the vector $\boldsymbol{u}_i$ associated with a particular λ_i can be found.

It is easy to show that the eigenvectors must be mutually orthogonal. Let $\boldsymbol{u}_r$, $\boldsymbol{u}_s$ be two different eigenvectors and let λ_r, λ_s be the corresponding eigenvalues. Then

$$\boldsymbol{A}\boldsymbol{u}_r = \lambda_r \boldsymbol{u}_r, \tag{A.27}$$

$$\boldsymbol{A}\boldsymbol{u}_s = \lambda_s \boldsymbol{u}_s. \tag{A.28}$$

Multiplying (A.27) by $\boldsymbol{u}_s{}^t$ and (A.28) by $\boldsymbol{u}_r{}^t$ gives

$$\boldsymbol{u}_s{}^t \boldsymbol{A}\boldsymbol{u}_r = \lambda_r \boldsymbol{u}_s{}^t \boldsymbol{u}_r, \tag{A.29}$$

$$\boldsymbol{u}_r{}^t \boldsymbol{A}\boldsymbol{u}_s = \lambda_s \boldsymbol{u}_r{}^t \boldsymbol{u}_s, \tag{A.30}$$

and since $\boldsymbol{A}$ is symmetric the left-hand sides of (A.29) and (A.30) are identical. It follows that $\lambda_r \boldsymbol{u}_s{}^t \boldsymbol{u}_r = \lambda_s \boldsymbol{u}_r{}^t \boldsymbol{u}_s$, and since by hypothesis $\lambda_r \neq \lambda_s$, $\boldsymbol{u}_r{}^t \boldsymbol{u}_s$ must be zero.

While the eigenvalues of a symmetric matrix are always real, the eigenvalues of a positive-definite symmetric matrix are in addition always positive. The proof is straightforward. If a general vector $\boldsymbol{x}$ is expressed in the coordinate system defined by the eigenvectors then $\boldsymbol{x} = \boldsymbol{U}^t\boldsymbol{x}'$ from (A.24). It follows that the quadratic form $\boldsymbol{x}^t\boldsymbol{A}\boldsymbol{x}$ becomes $\boldsymbol{x}'^t\boldsymbol{U}\boldsymbol{A}\boldsymbol{U}^t\boldsymbol{x}'$ when transformed into this coordinate system. Since the matrix $\boldsymbol{U}\boldsymbol{A}\boldsymbol{U}^t$ is equal to the diagonal matrix $\boldsymbol{A}'$ whose leading diagonal coefficients are the eigenvalues λ_i, this quadratic form reduces to

$$\sum_{i=1}^{n} \lambda_i (x_i')^2.$$

Hence the change of coordinates which reduces $\boldsymbol{A}$ to a purely diagonal matrix reduces the associated quadratic form to a sum of squares. If this quadratic form is always positive for any arbitrary non-zero vector $\boldsymbol{x}$ it follows that all the λ's must be positive.

There are a number of methods for obtaining the complete set of eigenvalues and eigenvectors of a given matrix which are more efficient from a computational point of view than the simple process outlined above. These are described in standard textbooks on numerical analysis and will not be dealt with here. However, there are also a number of processes for finding single eigenvalues. Two of these are of sufficient importance in engineering problems to justify a brief account here.

The first method is essentially one for finding the eigenvalue of largest modulus, this being also the algebraically largest eigenvalue in the case of a positive definite matrix. For any matrix $\boldsymbol{A}$ the effect of multiplying an eigenvector $\boldsymbol{u}$ by $\boldsymbol{A}$ is to multiply its "length" by the corresponding λ without changing its "direction". Any arbitrary vector $\boldsymbol{x}^{(0)}$ may be expressed in the coordinate system defined by the eigenvectors as

$$\boldsymbol{x}^{(0)} = x_1'\boldsymbol{u}_1 + x_2'\boldsymbol{u}_2 + \ldots + x_n'\boldsymbol{u}_n$$

where $x_1', x_2', \ldots, x_n'$ are the components of $\boldsymbol{x}^{(0)}$ in the directions of the corresponding $\boldsymbol{u}_i$'s. Multiplying $\boldsymbol{x}^{(0)}$ by $\boldsymbol{A}$ gives

$$\boldsymbol{x}^{(1)} = \boldsymbol{A}\boldsymbol{x}^{(0)} = \lambda_1 x_1'\boldsymbol{u}_1 + \lambda_2 x_2'\boldsymbol{u}_2 + \ldots \lambda_n x_n'\boldsymbol{u}_n$$

each component being multiplied by the appropriate eigenvalue. Suppose now that $\lambda_1 < \lambda_2 < \ldots < \lambda_n$. Then the component of $\boldsymbol{x}^{(0)}$ in the direction of $\boldsymbol{u}_n$ has been multiplied by a larger factor than any of the other components—in other words, the direction of $\boldsymbol{x}^{(1)}$ is closer to the direction of $\boldsymbol{u}_n$ than was the direction of $\boldsymbol{x}^{(0)}$. Repetition of the process gives

$$x^{(2)} = A^2x^{(0)} = \lambda_1{}^2x_1{}'u_1 + \ldots \lambda_n{}^2x_n{}'u_n$$

and with each successive multiplication the magnitude of the component of the vector $x^{(r)}$ in the direction of u_n increases relative to the magnitudes of the other components.† Eventually a stage is reached where $x^{(r)}$ is in approximately the same direction as u_n, the components associated with the other eigenvectors being negligible by comparison. Thus $x^{(r+1)} = Ax^{(r)} \approx \lambda_n x^{(r)}$, λ_n being obtained as the ratio of $|x^{(r+1)}|$ to $|x^{(r)}|$.

In the form described above the process will always converge to the *largest* eigenvalue, while in vibration problems it is usually the *smallest* one which is required. (The smallest eigenvalue corresponds to the lowest natural frequency of vibration, the eigenvector associated with this eigenvalue usually being called the "fundamental mode".) However, this difficulty is easily overcome. If a matrix A has eigenvalues λ_i and eigenvectors u_i then these satisfy the equation $(A - \lambda I)u = 0$. Multiplication of this equation by $-(1/\lambda)A^{-1}$ gives $(A^{-1} - (1/\lambda)I)u = 0$. It follows that the eigenvalues of A^{-1} are the reciprocals of those of A, while the eigenvectors are the same. Thus the smallest eigenvalue of A can be found by taking the reciprocal of the largest eigenvalue of A^{-1}. (It is assumed here that A is non-singular, so that $\lambda = 0$ is not an eigenvalue.)

The speed of convergence of this process clearly depends on the ratio of λ_n to the other λ's. It also depends on the choice of initial $x^{(0)}$—it is a considerable advantage if a vector is chosen which is already more or less "in the right direction". In vibration problems the process corresponds to assuming a mode of vibration which is then corrected, and it is usually possible to make a reasonable guess at this initial mode using physical intuition or model tests.

The second method is known as "Rayleigh's principle". This may be used to obtain an approximate value for an eigenvalue from an approximation to the corresponding eigenvector. As mentioned above, in a vibration problem it is usually possible to guess the approximate form of a mode of oscillation, particularly when the mode in question is the funda-

† The process is not concerned with the absolute lengths of the vectors $x^{(0)}, x^{(1)}, \ldots, x^{(r)}, \ldots$, but merely with the relative proportions of their components. In a numerical application of the method it is desirable to make the lengths approximately equal by applying a suitable scalar multiplier after each matrix multiplication.

mental, and in fact Rayleigh's principle is often used for estimating the lowest natural frequency of a system.

The method is extremely simple. If $\boldsymbol{u}_r$ is an eigenvector of $\boldsymbol{A}$, then by definition $\boldsymbol{A}\boldsymbol{u}_r = \lambda_r \boldsymbol{u}_r$. Pre-multiplying both sides of this equation by $\boldsymbol{u}_r^t$ gives $\boldsymbol{u}_r^t \boldsymbol{A} \boldsymbol{u}_r = \lambda_r \boldsymbol{u}_r^t \boldsymbol{u}_r$, or $\lambda_r = \boldsymbol{u}_r^t \boldsymbol{A} \boldsymbol{u}_r / \boldsymbol{u}_r^t \boldsymbol{u}_r$. If the vector $\boldsymbol{u}_r$ is only an approximation to the eigenvector then this equation will not be exact, but it can be shown that the error in the value of λ_r computed from this formula is of second order compared with the error in $\boldsymbol{u}_r$. Thus a relatively crude approximation to the eigenvector will often yield a satisfactory approximation to the corresponding eigenvalue. It can also be shown that if $\boldsymbol{u}_r$ is not an exact eigenvector then the calculated λ_r will lie between the greatest and the least of the λ's. It follows that when used to estimate the lowest natural frequency of a system Rayleigh's principle always gives a value which is somewhat greater than the true frequency.

In many vibration problems the equation which has to be solved is not $\boldsymbol{A}\boldsymbol{u} = \lambda \boldsymbol{u}$ but the more general one

$$\boldsymbol{A}\boldsymbol{u} = \lambda \boldsymbol{B}\boldsymbol{u} \tag{A.31}$$

where $\boldsymbol{A}$ and $\boldsymbol{B}$ are known symmetric matrices. It is possible to reduce (A.31) to the simpler form by pre-multiplying both sides by $\boldsymbol{B}^{-1}$, but this has the drawback that in general $\boldsymbol{B}^{-1}\boldsymbol{A}$ is not symmetric. The symmetry may be preserved by writing $\boldsymbol{B}$ as the product of two triangular matrices. It can be shown that if $\boldsymbol{B}$ is a positive-definite matrix then it is possible to express $\boldsymbol{B}$ in the form $\boldsymbol{L}^t\boldsymbol{L}$, where $\boldsymbol{L}$ is a lower triangular matrix (i.e. a matrix which has all its elements above the leading diagonal equal to zero). Thus (A.31) may be written as $\boldsymbol{A}\boldsymbol{u} = \lambda \boldsymbol{L}^t \boldsymbol{L} \boldsymbol{u}$. Introducing the vector $\boldsymbol{v} = \boldsymbol{L}\boldsymbol{u}$ allows this equation to be written as $\boldsymbol{A}\boldsymbol{L}^{-1}\boldsymbol{v} = \lambda \boldsymbol{L}^t \boldsymbol{v}$, or as

$$(\boldsymbol{L}^t)^{-1}\boldsymbol{A}\boldsymbol{L}^{-1}\boldsymbol{v} = \lambda \boldsymbol{v}. \tag{A.32}$$

The matrix $(\boldsymbol{L}^t)^{-1}\boldsymbol{A}\boldsymbol{L}^{-1}$ is symmetric if $\boldsymbol{A}$ is symmetric, but it should be noted that its eigenvectors $\boldsymbol{v}_i$ are not the same as those of $\boldsymbol{A}$. The $\boldsymbol{u}_i$'s must be obtained from the $\boldsymbol{v}_i$'s by applying the relationship $\boldsymbol{u}_i = \boldsymbol{L}^{-1}\boldsymbol{v}_i$.

When using Rayleigh's principle the more general form (A.31) gives no trouble. One merely pre-multiplies both sides of (A.31) by $\boldsymbol{u}_r$ in the same way as before, obtaining $\lambda_r = \boldsymbol{u}_r^t \boldsymbol{A} \boldsymbol{u}_r / \boldsymbol{u}_r^t \boldsymbol{B} \boldsymbol{u}_r$.

A.6. The solution of linear equations and the inversion of matrices

All computers are provided with sub-routines for carrying out the standard operations of matrix algebra, including the solution of linear equations and the inversion of matrices. The detailed design and construction of these routines is the province of a professional numerical analyst rather than a structural engineer. However, in most structural analyses carried out on a computer the bulk of the computer time is likely to be taken up by an equation-solving or inversion routine. It is desirable, therefore, for non-specialist computer users to have a general understanding of how these routines work.

Methods of solving sets of simultaneous equations may be divided into those which produce a sequence of approximations to the true solution, starting from arbitrary initial values of the variables, and those which obtain an exact solution (except for numerical rounding errors).

One of the earliest approximate methods to be developed was the Gauss–Seidel procedure. The equations are scanned repeatedly in a fixed order, equation i being used to obtain a new value of the variable x_i in terms of the current values of the other variables. The cycle is repeated until a sufficiently accurate set of answers has been obtained. The process converges provided that the matrix of the coefficients appearing in the equations is positive definite.

In many structural applications the simple Gauss–Seidel procedure converges too slowly to be acceptable. Better results can be achieved in manual calculations by using a procedure known as "relaxation". This also operates by making repeated adjustments to arbitrary initial values of the variables, but the adjustments are not made in any set order, the pattern of the calculation being determined by human intuition and experience rather than by the application of set rules. Sometimes, as in moment distribution, the method takes the form of a semi-physical procedure applied to the structure itself rather than to the linear equations describing it.

Many of the devices which the human analyst uses to accelerate the convergence of a relaxation process are non-systematic and therefore difficult to incorporate in a computer program. However, a method known as "systematic over-relaxation" has been used successfully in many computer applications, particularly with linear equations derived from finite difference or finite element problems. The method is well described by Varga (1962).

One of the earliest and still one of the most popular of the "exact" methods is Gaussian elimination. This method is capable of handling ill-conditioned systems of equations, and there is no need for the coefficient matrix to be either symmetric or positive definite. The method is described here in its simplest form—there are many variations of it, most of them designed to facilitate the arrangement of the coefficients in a compact tabular form when the calculation is done by hand.

Starting with the set of n linear equations

$$\begin{bmatrix} a_{11} & a_{12} & \cdot & \cdot & a_{1n} \\ a_{21} & a_{22} & \cdot & \cdot & a_{2n} \\ \cdot & \cdot & \cdot & \cdot & \cdot \\ \cdot & \cdot & \cdot & \cdot & \cdot \\ a_{n1} & a_{n2} & \cdot & \cdot & a_{nn} \end{bmatrix} \begin{bmatrix} x_1 \\ x_2 \\ \cdot \\ \cdot \\ x_n \end{bmatrix} = \begin{bmatrix} b_1 \\ b_2 \\ \cdot \\ \cdot \\ b_n \end{bmatrix}$$

the variable x_1 is eliminated from all but the first equation by subtracting suitable multiples of the first equation from the others. In terms of matrix coefficients this is done by adding $-a_{i1}/a_{11}$ times the first row of coefficients to the coefficients of row i and multiplying the constant b_1 by the same factor and adding it to the corresponding constant b_i, where i takes the successive values 2, . . ., n. This leaves the actual solution of the equations unchanged, and results in the original equations being replaced by

$$\begin{bmatrix} a_{11} & a_{12} & \cdot & \cdot & a_{1n} \\ 0 & a_{22}{}^* & \cdot & \cdot & a_{2n}{}^* \\ \cdot & \cdot & \cdot & \cdot & \cdot \\ \cdot & \cdot & \cdot & \cdot & \cdot \\ 0 & a_{n2}{}^* & \cdot & \cdot & a_{nn}{}^* \end{bmatrix} \begin{bmatrix} x_1 \\ x_2 \\ \cdot \\ \cdot \\ x_n \end{bmatrix} = \begin{bmatrix} b_1 \\ b_2{}^* \\ \cdot \\ \cdot \\ b_n{}^* \end{bmatrix}$$

where asterisks indicate altered coefficients. In the same way multiples of the second equation are subtracted from the third and following equations in such a way as to make the coefficients $a_{32}{}^*, \ldots, a_{n2}{}^*$ all zero, and the process is continued until all the coefficients below the leading diagonal have been reduced to zero. During this process the row used to alter the coefficients in the other rows is called the *pivotal row*, and its leading diagonal element is termed the *pivotal element*.

This procedure reduces the equations to the form

$$\begin{bmatrix} a_{11} & a_{12} & . & . & a_{1n} \\ 0 & a_{22}^* & . & . & a_{2n}^* \\ . & . & . & . & . \\ . & . & . & . & . \\ 0 & . & . & 0 & a_{nn}^* \end{bmatrix} \begin{bmatrix} x_1 \\ x_2 \\ . \\ . \\ x_n \end{bmatrix} = \begin{bmatrix} b_1 \\ b_2^* \\ . \\ . \\ b_n^* \end{bmatrix}$$

The value of x_n may now be found from the last equation, which gives $x_n = b_n^*/a_{nn}^*$. Substitution of this value into the $(n-1)$'th equation gives x_{n-1}, and the process can be continued until finally x_1 is found.

The total number of operations involved in this process is theoretically of order $n^3/3$. However, for structural applications this is usually a very pessimistic estimate, since the matrices which arise in structural calculations are generally both sparse and banded, i.e. they have a large number of zero elements, and the non-zero elements tend to lie near the leading diagonal. In these circumstances the number of operations can be made very much less than $n^3/3$.

The elimination method provides a simple check on the stability or otherwise of a structural system. As mentioned in Section A.2, the stability of a structure depends on the coefficients appearing in the load/displacement equations forming a positive-definite matrix. It can be shown that if all the leading diagonal elements $a_{11}, a_{22}^*, \ldots, a_{nn}^*$ of the triangular matrix formed by the elimination process are strictly positive then the original matrix is positive-definite The product of these leading diagonal elements is in fact equal to the value of the determinant of the original matrix, and the elimination method is probably the easiest way of computing this quantity if it is required.

If the initial matrix is symmetric it is easy to show that after the unknown x_1 has been eliminated from the last $n-1$ equations these equations are still symmetric Similarly, when the second column of zeros has been introduced the last $n-2$ equations are symmetric and so on. Thus the process may be carried out entirely in terms of the coefficients on the leading diagonal and above it, or alternatively the symmetry may be used to provide a check on the calculation.

So far it has been assumed that the coefficients in the equations are single numbers. Exactly the same process may be employed if they are sub-matrices. The only difference is the notational one associated with the row multipliers $-a_{i1}/a_{11}$, etc., which now have to be written in the form $-\boldsymbol{a}_{i1}\boldsymbol{a}_{11}^{-1}$, etc.

The purely sequential process described above is quite satisfactory for those structural calculations in which each pivotal element represents a direct stiffness or flexibility and is therefore non-zero. However, the process clearly fails if it encounters a zero pivot, and is likely to be inaccurate if one or more of the pivots are small compared with the other elements. This difficulty is easily avoided by a process known as exchange of pivots. The first column of the matrix is scanned to find the element in the column with the numerically largest value. The equations are then rearranged so that this element becomes a_{11}. It follows that all the multipliers $-a_{i1}/a_{11}$ are made less than unity, so that the process of modifying the rows of the matrix does not give rise to large numbers. The same operation is repeated with the second and subsequent columns, equations 2, . . ., n being rearranged so that a_{22}^* becomes the largest in the column $a_{22}^*, \ldots, a_{n2}^*$, etc. Thus throughout the reduction process the factors by which rows are multiplied before being added to other rows are always less than unity.

There is an important generalization of the basic step used in Gaussian elimination. This generalized step is defined as follows. In a set of linear equations $\boldsymbol{Ax} = \boldsymbol{b}$ let i define an equation (the pivotal equation) and j define a column of $\boldsymbol{A}$, where the coefficient a_{ij} (the pivotal coefficient) is non-zero. The step consists of the following two operations:

(a) Equation i is divided by a_{ij} (thus making the pivotal element unity).

(b) For $k = 1, \ldots, n$ $(k \neq i)$, a_{kj} times the modified pivotal equation is subtracted from equation k.

The result of these two operations is to reduce the equations

$$\begin{bmatrix} a_{11} & \cdot & \cdot & \cdot & \cdot & a_{1j} & \cdot & \cdot & \cdot & a_{1n} \\ \cdot & & & & & \cdot & & & & \cdot \\ \cdot & & & & & \cdot & & & & \cdot \\ a_{i1} & \cdot & \cdot & \cdot & \cdot & a_{ij} & \cdot & \cdot & \cdot & a_{in} \\ \cdot & & & & & \cdot & & & & \cdot \\ \cdot & & & & & \cdot & & & & \cdot \\ a_{n1} & \cdot & \cdot & \cdot & \cdot & a_{nj} & \cdot & \cdot & \cdot & a_{nn} \end{bmatrix} \begin{bmatrix} x_1 \\ \cdot \\ \cdot \\ \cdot \\ \cdot \\ \cdot \\ x_n \end{bmatrix} = \begin{bmatrix} b_1 \\ \cdot \\ \cdot \\ b_i \\ \cdot \\ \cdot \\ b_n \end{bmatrix}$$

to

$$\begin{bmatrix} a_{11}^* & \cdot & \cdot & \cdot & \cdot & 0 & \cdot & \cdot & \cdot & a_{1n}^* \\ \cdot & & & & & \cdot & & & & \cdot \\ \cdot & & & & & \cdot & & & & \cdot \\ a_{i1}^* & \cdot & \cdot & \cdot & \cdot & 1 & \cdot & \cdot & \cdot & a_{in}^* \\ \cdot & & & & & \cdot & & & & \cdot \\ \cdot & & & & & \cdot & & & & \cdot \\ a_{n1}^* & \cdot & \cdot & \cdot & \cdot & 0 & \cdot & \cdot & \cdot & a_{nn}^* \end{bmatrix} \begin{bmatrix} x_1 \\ \cdot \\ \cdot \\ \cdot \\ \cdot \\ \cdot \\ x_n \end{bmatrix} = \begin{bmatrix} b_1^* \\ \cdot \\ \cdot \\ b_i^* \\ \cdot \\ \cdot \\ b_n^* \end{bmatrix}$$

with column j reduced to zeros except for a single 1 in row i. Column j is termed a *reduced* column.

Exactly the same procedure may be applied to the matrix equation $\boldsymbol{Ax} = \boldsymbol{By}$, the pivotal element being selected from either $\boldsymbol{A}$ or $\boldsymbol{B}$. Provided the rows of $\boldsymbol{A}$ and $\boldsymbol{B}$ are operated on in a similar manner the essential relationship between $\boldsymbol{x}$ and $\boldsymbol{y}$ is not altered by the process.

This transformation does not appear to have an accepted name. We shall call it a Gauss–Jordan step in view of the method of matrix inversion with which it is commonly linked. However, it can also be applied to rectangular matrices, and forms the basis of the computational procedures described in Sections 7.7 and 8.5 of this book.

It is clear that n Gauss–Jordan steps with $i = j = 1, \ldots, n$ will reduce a square matrix $\boldsymbol{A}$ to a unit matrix and thereby produce the solution of the set of linear equations $\boldsymbol{Ax} = \boldsymbol{b}$. The same procedure applied to the equations $\boldsymbol{Ax} = \boldsymbol{Iy}$ will convert these equations into $\boldsymbol{Ix} = \boldsymbol{A}^{-1}\boldsymbol{y}$. Thus we have a systematic procedure for finding the inverse of a matrix.

As in Gaussian elimination, the selection of pivots from the leading diagonal (which is implied by making $i = j$) causes this process to fail if a zero leading diagonal element is encountered. While it is possible to avoid this by the pivotal exchange scheme described above, the same effect may be achieved more efficiently by selecting the pivotal rows sequentially, but choosing the numerically largest element in the row as pivot at each step. The procedure is illustrated by the inversion of the 3×3 matrix

$$\begin{bmatrix} 2 & 3 & -1 \\ 4 & 1 & 3 \\ 1 & 2 & 4 \end{bmatrix}$$

The corresponding linear relationship is

$$\begin{bmatrix} 2 & 3 & -1 \\ 4 & 1 & 3 \\ 1 & 2 & 4 \end{bmatrix} \begin{bmatrix} x_1 \\ x_2 \\ x_3 \end{bmatrix} = \begin{bmatrix} 1 & 0 & 0 \\ 0 & 1 & 0 \\ 0 & 0 & 1 \end{bmatrix} \begin{bmatrix} y_1 \\ y_2 \\ y_3 \end{bmatrix}$$

Taking the first row as the pivotal row it will be seen that the numerically largest element is in the second column. A Gauss–Jordan step is therefore carried out with $i = 1, j = 2$. This gives

$$\begin{bmatrix} 2/3 & \textcircled{1} & -1/3 \\ 10/3 & 0 & 10/3 \\ -1/3 & 0 & 14/3 \end{bmatrix} \begin{bmatrix} x_1 \\ x_2 \\ x_3 \end{bmatrix} = \begin{bmatrix} 1/3 & 0 & 0 \\ -1/3 & 1 & 0 \\ -2/3 & 0 & 1 \end{bmatrix} \begin{bmatrix} y_1 \\ y_2 \\ y_3 \end{bmatrix}$$

The pivot for the second equation may be chosen arbitrarily from the first or third column of A. Applying a Gauss–Jordan step with $i = 2$, $j = 3$ gives

$$\begin{bmatrix} 1 & 1 & 0 \\ 1 & 0 & \textcircled{1} \\ -5 & 0 & 0 \end{bmatrix} \begin{bmatrix} x_1 \\ x_2 \\ x_3 \end{bmatrix} = \begin{bmatrix} 3/10 & 1/10 & 0 \\ -1/10 & 3/10 & 0 \\ -2/10 & -14/10 & 1 \end{bmatrix} \begin{bmatrix} y_1 \\ y_2 \\ y_3 \end{bmatrix}$$

For the last row there is no choice of pivot. A final Gauss–Jordan step with $i = 3, j = 1$ gives

$$\begin{bmatrix} 0 & 1 & 0 \\ 0 & 0 & 1 \\ \textcircled{1} & 0 & 0 \end{bmatrix} \begin{bmatrix} x_1 \\ x_2 \\ x_3 \end{bmatrix} = \begin{bmatrix} 13/50 & -9/50 & 1/5 \\ -7/50 & 1/50 & 1/5 \\ 2/50 & 14/50 & -1/5 \end{bmatrix} \begin{bmatrix} y_1 \\ y_2 \\ y_3 \end{bmatrix}$$

Finally, rearranging the equations gives

$$\begin{bmatrix}1 & 0 & 0\\0 & 1 & 0\\0 & 0 & 1\end{bmatrix}\begin{bmatrix}x_1\\x_2\\x_3\end{bmatrix}=\frac{1}{50}\begin{bmatrix}2 & 14 & -10\\13 & -9 & 10\\-7 & 1 & 10\end{bmatrix}\begin{bmatrix}y_1\\y_2\\y_3\end{bmatrix}$$

As a result of these operations $\boldsymbol{A}$ has been reduced to $\boldsymbol{I}$ and $\boldsymbol{I}$ has been replaced by $\boldsymbol{A}^{-1}$. It is easy to verify by direct multiplication that

$$\begin{bmatrix}2 & 3 & -1\\4 & 1 & 3\\1 & 2 & 4\end{bmatrix}\times\frac{1}{50}\begin{bmatrix}2 & 14 & -10\\13 & -9 & 10\\-7 & 1 & 10\end{bmatrix}=\begin{bmatrix}1 & 0 & 0\\0 & 1 & 0\\0 & 0 & 1\end{bmatrix}$$

In this example the vectors $\boldsymbol{x}$ and $\boldsymbol{y}$ have been included for the sake of clarity. From a computational point of view they are irrelevant, since all the operations are carried out solely on the elements of the two matrices. It may be noted that at any stage of the process there are always three columns (or n columns in the general case) which contain only a single 1. In a computer implementation of the process it is possible to avoid storing these reduced columns, so that a matrix can be inverted by this method without any additional storage being required. The number of operations required is of the order of n^3, i.e. three times that required for the solution of the corresponding set of linear equations.

A.7. The linear programming problem

This problem first arose in operational research, and the first formal procedures for solving it were constructed in the mid-1940s. The problem may be stated as follows:

> "Find the values of a set of n non-negative variables† x_j ($j = 1, \ldots, n$) which maximize (or minimize) a linear function $w = c_j x_j$ and satisfy a set of m linear constraints $a_{ij}x_j >, =$ or $< b_i$ ($i = 1, \ldots, m$)."

† The condition that the variables must be non-negative was a natural one in the context of the origination of the problem, and forms an integral part of many of the numerical solution procedures. It is not, however, a serious restriction. Problems involving variables of unrestricted sign can always be converted to "standard form" by writing each variable as the difference of two positive variables.

The nature of the problem and its solution may be seen from the following simple two-dimensional example:

$$\text{Maximize } w = x_1 + x_2, \text{ subject to} \quad \begin{aligned} 3x_1 + 5x_2 &\leqslant 40 \\ x_1 + x_2 &\geqslant 3 \\ 0 \leqslant x_1 &\leqslant 7 \\ 0 \leqslant x_2 &\leqslant 5 \end{aligned}$$

These inequalities define a "permissible region" in a space with coordinate axes x_1, x_2, as shown in Fig. A.3. Points belonging to this region are termed "feasible solutions". Since w is a linear function of the variables the contours of constant w are straight lines, and the maximum is obviously attained at a vertex of the permissible region, in this case the point P. Thus only two of the inequalities are necessary to define the solution and these two are satisfied as equalities.

The form of the solution is not greatly altered if some of the constraints are equalities. In this example an equality constraint reduces the permissible region to a line segment. Thus if the first inequality is replaced by the equality $3x_1 + 5x_2 = 40$ the permissible region is reduced to the line segment PQ. The actual solution, however, is unchanged.

The characteristics of the general n-variable problem are very similar. The permissible region becomes a convex polyhedron in a linear vector space with the n variables as coordinate axes, while the surfaces of constant w form a set of parallel hyperplanes. The solution is always a vertex of the permissible region, at which n of the constraints are satisfied as equalities.

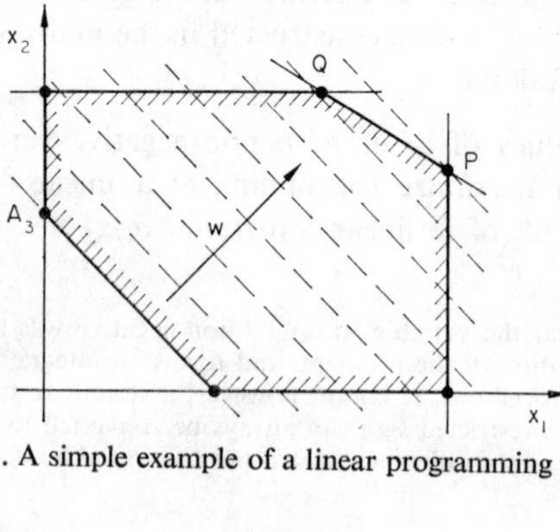

FIG. A.3. A simple example of a linear programming problem.

It follows that the solution of a linear programming problem involves solving the n linear equations which define the vertex of the permissible region at which w is a maximum (or minimum). This requires some procedure for selecting the right set of equations from the complete set of constraints. (One might, of course, solve all the sets of n equations and compare the associated values of w, but this is clearly an extremely inefficient process.) The earliest of such procedures to be developed was Dantzig's Simplex method. Like most of its successors this first finds an arbitrary vertex of the permissible region (known as a "basic feasible solution") and then looks for an adjacent vertex with a larger value of w. In this way it reaches the correct solution after examining only a small fraction of the complete set of vertices. Details of the Simplex and other methods will be found in the standard texts such as Dantzig (1963) and Hadley (1962).

There is a result in linear programming theory known as the duality theorem which has close links with the upper and lower bound theorems of plasticity. The two linear programming problems

$$\text{"Maximize } w = \boldsymbol{c}^t\boldsymbol{x} \text{ subject to } \boldsymbol{A}\boldsymbol{x} \leqslant \boldsymbol{b},\ \boldsymbol{x} \geqslant \boldsymbol{0}\text{"}$$

and

$$\text{"Minimize } \bar{w} = \boldsymbol{b}^t\boldsymbol{y} \text{ subject to } \boldsymbol{A}^t\boldsymbol{y} \geqslant \boldsymbol{c},\ \boldsymbol{y} \geqslant \boldsymbol{0}\text{"}$$

are said to be dual problems. If the first problem (sometimes called the primal) has n variables and m constraints then the second problem (the dual) has m variables and n constraints. The theorem of duality states that if one of these problems has a finite solution then so has the other and that the values of w and $\bar{w}$ associated with these solutions are equal. This result is referred to in Section 7.4.

References

(a) *General books on matrix methods of structural analysis*

PRZEMIENIECKI, J. S. (1968) *Theory of Matrix Structural Analysis.* McGraw-Hill.
This book covers both continuum and skeletal structures and contains a comprehensive bibliography.

ZIENKIEWICZ, O. C. (1971) *The Finite Element Method in Engineering Science*, 2nd ed. McGraw-Hill.
This book is almost entirely concerned with the solution of continuum problems.

(b) *General books on matrix algebra and associated computing techniques*

BICKLEY, W. G. and THOMPSON, R. S. H. G. (1964) *Matrices: Their Meaning and Manipulation.* E.U.P.

LIEBECK, H. (1969) *Algebra for Scientists and Engineers.* Wiley.

(c) *References quoted in the text of this book*

ARCHER, J. S. (1963) *Proc. Amer. Soc. Civil Engineers*, **89,** ST4, 161.

BAKER, Sir JOHN F. and HEYMAN, J. (1969) *Plastic Design of Frames.* C.U.P.

CLOUGH, R. W., WILSON, E. L. and KING, I. P. (1963) *Proc. Amer. Soc. Civil Engineers*, **89,** ST4, 179.

DANTZIG, G. B. (1963) *Linear Programming and Extensions.* Princeton University Press.

FOX, R. L. (1971) *Optimization Methods for Engineering Design.* Addison-Wesley.

GALLAGHER, R. H. and ZIENKIEWICZ, O. C. (Editors) (1973) *Optimum Structural Design.* Wiley.

HADLEY, G. (1962) *Linear Programming*. Addison-Wesley.

HENDERSON, J. C. DE C. and BICKLEY, W. G. (1955) *Aircraft Engineering*, **27,** 400.

HEYMAN, J. (1958) *Proc. Amer. Soc. Civil Engineers*, **84,** EM4, paper no. 1790.

HORNE, M. R. and MERCHANT, W. (1965) *The Stability of Frames.* Pergamon Press.

IRONS, B. M. (1970) *Int. J. Num. Methods in Engineering*, **2,** 5.

KRON, G. (1944) *J. Franklin Inst.* **238,** 6, 400.

LECKIE, F. L. and LINDBERG, G. M. (1963) *The Aeronautical Quart.* **14,** 224.

LIVESLEY, R. K. and CHANDLER, D. B. (1956) *Stability Functions for Structural Frameworks.* Manchester University Press.

LIVESLEY, R. K. (1973) Chapter 6 of *Optimum Structural Design* (Ed. Gallagher, R. H. and Zienkiewicz, O. C.). Wiley.

LUNDQUIST, E. E. and KROLL, W. D. (1944) N.A.C.A. Report ARR no. 4B24.

MORICE, P. B. (1959) *Linear Structural Analysis*, Chap. 3. Thames and Hudson.
NEAL, B. G. (2nd ed. 1963) *The Plastic Methods of Structural Analysis*. Chapman and Hall.
OSTENFELD, A. (1926) *Die Deformations Methode*. Springer, Berlin.
PESTEL, E. C. and LECKIE, F. L. (1963) *Matrix Methods in Elasto-mechanics*. McGraw-Hill.
ROBINSON, J. (1966) *Structural Matrix Analysis for the Engineer*. Wiley.
TEWARSON, R. P. (1973) *Sparse Matrices*. Academic Press.
VARGA, R. S. (1962) *Matrix Iterative Analysis*. Prentice-Hall.

Index